P9-BBU-038

LABORATORY ANIMAL ANAESTHESIA

SECOND EDITION

A practical introduction for research workers and technicians

LABORATORY ANIMAL ANAESTHESIA

SECOND EDITION

A practical introduction for research workers and technicians

P. A. Flecknell

Comparative Biology Centre
The Medical School
Newcastle-Upon-Tyne UK

ELSEVIER
BUTTERWORTH
HEINEMANN

AMSTERDAM BOSTON HEIDELBERG LONDON NEW YORK OXFORD
PARIS SAN DIEGO SAN FRANCISCO SINGAPORE SYDNEY TOKYO

This book is printed on acid-free paper

First edition published in 1987 by Academic Press
Reprinted 2000, 2003

Elsevier Academic Press
525 B Street, Suite 1900, San Diego, California 92101-4495, USA
http://www.elsevier.com

Elsevier Academic Press
84 Theobald's Road, London WC1X 8RR, UK
http://www.elsevier.com

British Library Cataloguing in Publication Data
A catalogue record for this book is available from the British Library

ISBN 0-12-260361-3

Typeset by J & L Composition Ltd, Filey, North Yorkshire
Printed and bound in Malta by Gutenberg Press
03 04 05 06 07 08 9

Contents

Preface

The majority of laboratory animals are anaesthetized by staff who have not received specialist training in this field. Unfortunately, most textbooks of human or veterinary anaesthesia assume that the reader has a basic knowledge of the subject. Because of this, a good deal of published information has remained relatively inaccessible and this has limited the introduction of new techniques into the field of laboratory animal anaesthesia.

This handbook attempts to provide a basic guide to anaesthesia for research workers and animal technicians. It is not intended to be a comprehensive text on animal anaesthesia, but concentrates on those areas that are of greatest practical importance when anaesthetizing laboratory animals.

The first sections of the book deal with the general principles of pre-operative care, anaesthetic techniques and anaesthetic management. The most important properties of the anaesthetic and other agents used are outlined, but a detailed description of their pharmacology has been deliberately excluded. These sections also provide details of some of the equipment that the author has found useful when anaesthetizing laboratory animals.

These general sections of the book should be read before using any of the anaesthetic regimens described in the final sections. In particular, it is hoped that the reader will study the sections on post-operative care and the provision of effective pain relief before carrying out any operative procedures on animals.

In order to provide rapid, easily accessible guidelines a list of recommended anaesthetic regimens for each of the common laboratory species is given in Appendix 1. For research workers who require alternative techniques, a wider range of anaesthetic regimens is discussed together with an extensive list of dose rates for each species in Chapter 7.

In addition to providing guidance on basic anaesthetic technique, an introduction to more specialist procedures such as long-term anaesthesia and the use of neuromuscular blocking agents has been included. These sections provide only initial guidance and it is recommended that, whenever possible, an experienced veterinary anaesthetist should be consulted before attempting these techniques.

P. A. FLECKNELL

Preface to the second edition

Since writing the first edition of this book there has been a welcome increase in concern for the welfare of laboratory animals. One result of this has been the introduction by a number of countries of formal training requirements for new research workers. This increased interest in animal welfare has also led to the improved dissemination of information regarding 'best practice' in many aspects of laboratory animal science. The second edition of *Laboratory Animal Anaesthesia* has benefited from this exchange of information, and the additions and revisions which have been included owe much to comments from my colleagues around the world. A major addition to this new edition is the inclusion of illustrations of techniques and equipment. The format of the book remains relatively unchanged, except for Chapter 7, which now incorporates some of the information previously included in the Appendices. This enables more of the information relating to a particular species to be accessed quickly and easily. Brief descriptions of anaesthetic techniques for fish, amphibia, reptiles and birds have also been included, to provide some guidance for dealing with these species.

P. A. FLECKNELL

Glossary

Inevitably, a number of specialist terms are used throughout this book and these are defined below.

Anaesthesia	a state of controllable, reversible insensibility in which sensory perception and motor responses are both markedly depressed
Analgesia	the temporary abolition or diminution of pain perception
Analeptic	drug which stimulates respiration
Anoxia	complete deprivation of oxygen for tissue respiration
Apnoea	temporary cessation of breathing
Arrhythmia	(cardiac) arrhythmias are alterations in the normal rhythm of the heart
Asystole	lack of cardiac muscle contractions
Ataxia	lack of co-ordination, 'wobbliness'
BMR	basal metabolic rate
Bradycardia	slowing of the heart rate
CNS	central nervous system
CNS depressant	any agent that modifies function by depressing sensory or motor responses in the CNS

Cyanosis	blue or purple colouring of the skin or visible membranes due to the presence of an increased concentration of reduced haemoglobin in capillary blood, symptomatic of hypoxia.
Dosages	all dosages are expressed as milligrams of drug per kilogram of body weight (mg kg^{-1}), except for the neuroleptanalgesic combinations which are more conveniently expressed as ml of commercial or diluted premixed solution per kg body weight (ml kg^{-1})
Dosage schedules	u.i.d. — once daily b.i.d. — twice daily t.i.d. — three times daily q.i.d. — four times daily
Dyspnoea	laboured breathing
ECG	electrocardiogram
Hypercapnia	elevated blood carbon dioxide content
Hyperpnoea	fast or deep breathing
Hypertension	elevated (arterial) blood pressure
Hypnotic	a drug that induces a state resembling deep sleep, but usually with little analgesic effect
Hypocapnia	reduced blood carbon dioxide content
Hypopnoea	slow or shallow breathing
Hypotension	a fall in (arterial) blood pressure
Hypothermia	a fall in body temperature
Hypovolaemia	a fall in circulating blood volume

Hypoxia	depressed levels of oxygen
Induction (of anaesthesia)	the initial establishment of a state of anaesthesia
Injections	routes of administration are abbreviated as follows: i/v — intravenous i/m — intramuscular i/p — intraperitioneal s/c — subcutaneous
Laryngospasm	spasm of the vocal cords, producing complete or partial obstruction of the airway
Minute volume	the volume of gas breathed in 1 minute, i.e. the product of tidal volume and respiratory rate
Narcosis	a state of insensibility or stupor from which it is difficult to arouse the animal
Normovolaemic	having a normal circulating blood volume
$P\text{CO}_2$	partial pressure of carbon dioxide
per os	by mouth
$P\text{O}_2$	partial pressure of oxygen
Polypnoea	rapid, panting breathing
Pulmonary ventilation	the mechanical expansion and contraction of the lungs in order to renew alveolar air with fresh atmospheric air
Tachycardia	an increase in heart rate
Tachypnoea	rapid respiration
Tidal volume	the volume of gas expired with each breath

List of Tables

List of Figures

Acknowledgements

I would like to thank my colleagues at the Clinical Research Centre, Harrow and the University of Newcastle upon Tyne for their advice and helpful criticism of various parts of this text. In particular, I am grateful to Dr Richard Wootton for his many helpful comments during the development of this book.

Thanks are also due to Anne Al-Jumaili and Maggie Smith for typing the manuscript and for coping with the numerous revisions of the text so cheerfully.

I am particularly grateful to Ruth, my wife, for her constant assistance during all stages of the preparation of this book.

During the preparation of the second edition, many colleagues and friends have provided helpful advice and comments, in particular Julie Drage, Anna Meredith, Tim Morris, Peter Nolan and Simon Young. Special thanks are due to Ken Boschert, the instigator and operator of COMPMED. Without the benefit of this electronic conferencing system, the new edition of this book could not have benefited from the comments, contributions and requests for additional information from the many 'COMPMEDDERs' involved in anaesthetizing laboratory animals around the world.

Finally, I wish to thank my colleagues in the Audio-Visual Centre at Newcastle for providing the illustrations for this new edition.

Introduction

The use of safe and effective anaesthetic techniques can have a major influence on the welfare of laboratory animals. Improvement of anaesthetic techniques should be considered an essential aspect of the refinement of experimental methods advocated by Russell and Burch (1992). One of the main concerns voiced by the general public is that if animals must be used for experimental purposes, pain and distress should be abolished or reduced to an absolute minimum. The provision of effective anaesthesia and analgesia therefore addresses one of the fundamental public concerns with regard to animal experimentation. It is important to appreciate that poor anaesthetic technique can have a serious adverse effect on the quality of results obtained from animal experimentation. Clearly, it is the responsibility of research workers to review their current anaesthetic practices and to introduce improvements whenever possible. Along with a review of anaesthetic techniques, attention should also be given to the pre-operative and post-operative care of the animal. A common failing of current laboratory anaesthetic practice is the lack of consideration given to this aspect of animal care. Attention to the needs and well-being of the animal both before and after a surgical procedure must be considered an integral part of the anaesthetist's responsibilities. The provision of effective pain relief post-operatively is of particular importance, and is discussed in detail in Chapter 6.

1

Pre-operative Care

To provide anaesthesia of the standard required in modern research laboratories, it is essential that adequate pre-operative preparations are made before attempting to anaesthetize an animal. Good pre-operative care will reduce the incidence of many of the complications that can occur during anaesthesia, and thorough preparation of facilities and equipment also contributes to the smooth running of a research protocol. It is important to consider preparation not only of the animals that are to be anaesthetized, but also the equipment, drugs, facilities and personnel that will be involved in the procedure.

I. ANAESTHETIC EQUIPMENT, ANAESTHETIC DRUGS AND PERSONNEL

Having determined an anaesthetic protocol, it is important to establish that all the equipment necessary is available, and in good working order. A simple pre-use check list for anaesthetic machines is given in Table 1.1. Ensure that sufficient anaesthetic drugs and anaesthetic gases have been provided not just for the anticipated period of anaesthesia, but to cover unexpected additional requirements (see Appendix 2 for estimations of quantities of inhaled anaesthetic agents). Check the expiry date of all drugs, and that they have been properly stored. In addition to the anaesthetic agents, drugs needed for coping with emergencies must be readily available (see Chapter 4).

Monitoring equipment should be switched on, allowed a period to stabilize if necessary, and its functions checked. Alarm limits should be reset from the default settings (which are often values appropriate for human patients), and then fine-tuned when the individual animal is connected. Heating pads and blankets should be switched on approximately 30–60 minutes before they are needed, to allow them to attain the desired operating temperature.

Finally, if the animal is to recover from anaesthesia, check that a suitable

TABLE 1.1
Pre-anaesthetic checks of anaesthetic equipment. These pre-anaesthetic checks should become routine procedures and will minimize the occurrence of anaesthetic accidents which could result in the death of the animal

- Is only one oxygen cyclinder in use and the other full?
- Check that the valve on the cylinder which is in use is opened fully to provide a free flow of gas (the reading on the pressure dial on an oxygen cylinder gives a reasonable indication as to how much oxygen it contains)
- Check that the cylinders are full and properly attached to the anaesthetic machine, ensure the flowmeters are functioning correctly by opening the cylinder valves and opening the needle valves that control the flow of gas through the flowmeters. The bobbins should rotate when gas is flowing (most are marked with a small white dot to assist in assessing this). The gas flow rate is measured from the top of the bobbin. Turn off the gas flow using the needle valve and check that the bobbin sinks smoothly back to zero, and is not sticking and giving a false high gas flow rate
- Check that the emergency oxygen switch on the machine is functioning correctly
- If an anaesthetic circuit with valves is to be used, open them fully and check that they are operating correctly
- If a volatile anaesthetic is to be used, check that the vaporizer has been filled and that the control dial moves smoothly over the entire range of possible settings. If using a machine with several vaporizers, check that the correct one has been selected. Check that when the vaporizer is turned off, no anaesthetic odour can be detected when the oxygen flowmeter is turned on
- If the anaesthetic machine has a built-in circle type absorber ensure that this is switched out of circuit (usually marked 'open') if the absorber is not to be used. Check that the soda lime has not become exhausted (indicated by a colour change from pink to white or white to violet)
- Attach the circuit which will be used to the anaesthetic machine, turn on the oxygen supply and check the circuit for leaks by occluding the patient end of the tubing and fully closing any valves. Open the valves to check they are not sticking
- If a mechanical ventilator is to be used, switch it on and observe it for a few respiratory cycles. If possible, check the tidal volume that is being delivered with a respirometer
- Run through the manufacturer's recommended pre-use check on any monitoring equipment

area for post-operative recovery has been provided (see Chapter 6), and that any incubators or heat pads needed in the recovery period are switched on well in advance.

If personnel have been allocated to assist with anaesthesia, check that they have been properly briefed about the research protocol, and are familiar with the equipment and techniques that will be used. Ensure that they are aware of the time they are required, including attendance for post-operative observation and care, which may be outside the normal working day.

II. THE ANIMAL

The single most important factor that can reduce the risks associated with anaesthesia is the use of healthy animals. It is most important to ensure that any animals that are to be anaesthetized are at least in overt good health and free from subclinical disease. Whenever possible, animals of defined health status should be obtained, so that the occurrence of respiratory and other diseases can be eliminated. Anaesthetizing animals that have intercurrent infections, even if these are causing no overt signs of disease, usually results in increased mortality and morbidity. In addition to the wasted resources and animal welfare implications, intercurrent disease increases variability in research data and so requires use of larger numbers of animals.

A. Acclimatization

Animals should be obtained at least 7 days and preferably 14 days prior to their intended use, so that an appropriate period is allowed for acclimatization to their new environment. Requirements vary in different establishments, and research workers should check on local practices. During this period the metabolic and hormonal changes caused by the stress of transportation will return to normal, and the animal can be monitored for any signs of ill-health. Animal care staff and research workers will have the opportunity to familiarize themselves with the behaviour and particular characteristics of the group of animals, and records of body weight, growth rate, and food and water consumption can be commenced. This information is invaluable if animals are intended to recover from anaesthesia after undergoing a surgical procedure. Many of the pain assessment schemes which are under development rely on a knowledge of these variables, and it is important that such information is obtained and recorded (see Chapter 6). Even when planning non-recovery procedures, an assessment of food and water intake or growth rate will provide some reassurance that the animal is in a normal physiological state.

Acclimatization of species that can rapidly develop a relationship with their handler (e.g. dogs, cats and pigs) has the advantage of reducing unnecessary distress during induction and recovery from anaesthesia. Regular handling of most species, including small rodents, will habituate the animals to the procedure. As a consequence, the animals will be easier to restrain, more co-operative, and induction of anaesthesia will be safer both for the animal and the staff involved.

B. Clinicial examination

Whatever the health status of the animal, it is useful to carry out some form of clinical examination before induction of anaesthesia. Although many investigators may not be familiar with diseased animals they are often very familiar with normal animal behaviour and appearance. If there is any deviation from the normal, further advice can be sought from animal technicians and veterinarians. The presence of discharges from the eyes or nose, matting of the fur around these regions, or soiling of the perianal region with faeces all require further investigation. If the overall appearance of the animal is abnormal, or any of the clinical signs mentioned are present, induction of anaesthesia should be delayed until expert advice can be obtained. As mentioned above, it is helpful to monitor food and water intake and body weight for a few days pre-operatively. This will allow an assessment to be made of whether the intake is normal and will also be of use in monitoring the post-operative recovery of the animal.

C. Pre-operative fasting

Cats, dogs, ferrets, primates and pigs should receive no food during the 8–12 hours before anaesthesia, in order to minimize the risk of vomiting during induction of anaesthesia or during recovery. Withholding food from ruminants has virtually no effect on the volume of digesta which remain in the rumen, unless excessive periods of starvation are employed (3–4 days), but a short period of starvation (12–24 hours) may help to reduce the incidence of ruminal tympany or bloat (the accumulation of gas in the stomach).

Pre-anaesthetic fasting of rabbits and small rodents is unnecessary since vomiting during induction does not occur in these species. Problems may occasionally be seen with guinea pigs which may retain food in their pharynx after being anaesthetized. If this occurs in a significant number of animals then a short period of pre-anaesthetic fasting (6–8 hours) should be introduced. If gastrointestinal tract surgery is to undertaken, then fasting may be required in all species, but it is important to note that rodents and rabbits are coprophagic, so measures to prevent them ingesting their faeces may be necessary in order to provide a completely empty stomach. An additional complication arises because of the diurnal rhythms of these species. Although food may be provided immediately post-operatively, it may not be eaten until the onset of the dark phase of the animal's photoperiod. In addition, if food intake is depressed because of pain, surgical stress or delayed recovery from anaesthesia, food and water intake may be severely depressed for at least 24 hours post-operatively. The metabolic consequences of this, especially when coupled with pre-operative fasting, can be severe and can compromise both the research data obtained and animal welfare. It is therefore preferable to withhold food only when required by a particular research protocol. Withholding food from pregnant animals of all species, but in particular ruminants and guinea pigs, can produce severe metabolic disturbances which may prove fatal.

Large or medium-sized birds (e.g. ducks, chickens, and pigeons) may be starved for 6–12 hours to reduce the risk of regurgitation of the contents of the crop. Smaller birds should not fast for longer than 2 hours, to avoid the risk of inducing hypoglycaemia. Fasting of reptiles and amphibia is generally unnecessary. Fish should be starved for 24–48 hours prior to anaesthesia.

All animals should be provided with drinking water until approximately 60 minutes prior to induction of anaesthesia. If the animal has a reduced fluid intake, or if vomiting, diarrhoea or haemorrhage have occurred, then some pre-operative fluid therapy will be necessary. The basic principles are outlined in Chapter 5, but whenever possible veterinary advice should be obtained.

Whenever practicable, animals should be weighed prior to anaesthesia, both to allow accurate calculation of drug dosages and to enable assessment of any post-operative weight loss.

2

Pre-anaesthetic Medication

Pre-anaesthetic medication is given in order to:

- Reduce fear and apprehension, provide sedation and aid stress-free induction of anaesthesia.
- Reduce the amount of other anaesthetic agents required to induce general anaesthesia, so decreasing the undesirable side-effects of these agents.
- Provide smoother induction of anaesthesia.
- Provide smoother recovery from anaesthesia.
- Reduce the volume of salivary and bronchial secretions which might block the airways.
- Block the vaso-vagal reflex (the reflex slowing of the heart that can occur as a result of endotracheal intubation and surgical procedures).
- Reduce pre-operative pain and minimize pain in the immediate post-operative period.

Although the aims listed above apply to all animal species, pre-anaesthetic medication is often used in larger animals, where sedation and tranquilliza-tion are required to aid humane restraint, and minimize the risk of injury to the animal and its handler. Pre-anaesthetic medication should be used in a wider range of species, since even when restraint is not a problem, the use of sedatives and tranquillizers may be advantageous. In humans, many of the drugs used have been shown to reduce fear and allay anxiety, and similar effects may occur in animals.

In addition to the use of drugs, careful and expert handling of laboratory animals is an essential part of humane pre-anaesthetic management. Con-sideration of the techniques used and their possible stressful effects upon the animal should enable modification of procedures to minimize pain or distress. For example, administration of a sedative or analgesic to an animal still housed in its pen or cage, followed by removal to the operating theatre or research laboratory only after the drug has taken effect, can considerably reduce the stress that might otherwise be caused.

If an intravenous induction agent is to be used, then it is helpful to apply a local anaesthetic cream (EMLA, Astra) to the skin overlying the vein, about 45–60 minutes before intravenous injection. This eliminates the pain or discomfort of venepuncture, and has the added advantage of eliminating any movement in response to the procedure, since the skin is completely anaesthetized (Flecknell *et al.*, 1990a).

The selection of a pre-anaesthetic drug regimen will depend on the species of animal to be anaesthetized; the anaesthetic agents to be used; the particular requirements of the research protocol; and the personal preferences of the anaesthetist. The characteristics of the major groups of drugs available are listed below and detailed recommendations for each species are given in Chapter 7.

I. ANTICHOLINERGIC DRUGS

A. Atropine

1. Desirable effects

Desirable effects include reduction of bronchial and salivary secretions which might partially occlude the airways. Atropine protects the heart from vagal inhibition, which can occur during endotracheal intubation or during surgical procedures, particularly if the viscera are handled. Atropine may also be used to correct any slowing of the heart caused by opioids such as fentanyl.

2. Undesirable effects

Undesirable effects include increased heart rate. In ruminants, atropine does not completely block salivary secretions, which become more viscous.

3. Special comments

Avoid the use of atropine if the heart rate is already elevated; also avoid if tachycardias are likely to be produced (e.g. during cardiac surgery). Atropine is rapidly metabolized in some strains of rabbits and so its effects may be unpredictable in this species.

B. Glycopyrrolate

1. Desirable effects

Reduction of salivary and bronchial secretions, protection of the heart from vagal inhibition.

2. Undesirable effects

Increased heart rate, although less pronounced than with atropine in some species.

3. Special comments

Glycopyrrolate has a longer duration of action than atropine and has been reported to be the more effective agent in rodents and rabbits (Olson *et al.*, 1993) It is the anticholinergic agent of choice in rabbits because its duration of action is less affected by the high levels of atropinase that may be present in this species. Glycopyrrolate does not cross the blood–brain barrier, and in humans produces less visual disturbances than atropine. This may be advantageous in some animal species.

II. TRANQUILLIZERS AND SEDATIVES

Tranquillizers produce a calming effect without causing sedation. At high doses they produce ataxia and depression, but animals are readily roused, particularly in response to painful stimuli, since these drugs have no analgesic properties. Sedatives produce drowsiness and appear to reduce fear and apprehension in animals. There is considerable overlap in the action of many agents and a good deal of species variation in their effects, making definitive classification of drugs as sedatives or tranquillizers difficult.

A. Phenothiazines: chlorpromazine, acepromazine, promazine

1. Desirable effects

These drugs produce sedation, potentiate the action of anaesthetics, hypnotics and narcotic analgesics, and so reduce the dosage required to produce surgical anaesthesia. Sedation may extend into the post-operative period, so that recovery from anaesthesia is smooth.

2. Undesirable effects

A moderate hypotension may occur because of the production of peripheral vasodilation. Temperature regulation is depressed and moderate hypothermia may occur.

3. Special comments

The undesirable effects noted above are well tolerated by normal animals, but the drugs should not be used in animals with any form of fluid deficit, e.g. dehydration or haemorrhage. This group of drugs have no analgesic action but they potentiate the action of opiates.

B. Butyrophenones: droperidol, fluanisone, azaperone

1. Desirable effects

These drugs have effects similar to the phenothiazines (above), but are more potent.

2. Undesirable effects

The hypotensive effects of these drugs are generally less severe than those caused by phenothiazines.

3. Special comments

Butyrophenones such as droperidol and fluanisone are most widely used as components of neuroleptanalgesic combinations (see Chapter 3).

C. Benzodiazepines: diazepam, midazolam

1. Desirable effects

Desirable effects include sedation, but there is considerable species variation in effect: sedation is minimal in dogs, but marked in rabbits and rodents. Benzodiazepines potentiate the action of most anaesthetics and narcotic analgesics. They produce good skeletal muscle relaxation (NB: *not* muscle paralysis). A specific antagonist, flumazenil, is available, so that sedation can be reversed if necessary (Pieri *et al.*, 1981; Amrein and Hetzel, 1990).

2. Undesirable effects

In some species (dog and cat) benzodiazepines may cause mild excitement and disorientation rather than sedation. Injection of some preparations of diazepam into small blood vessels can cause irritation and damage to the vessel.

3. Special comments

Benzodiazepines (e.g. diazepam, midazolam) have both potent tranquillizing and sedative actions. Diazepam is the agent most frequently used, although some injectable formulations in organic solvents cannot be mixed with other water-soluble agents. Midazolam has effects similar to diazepam but with a shorter duration of action. Unlike diazepam, it is water-soluble, and so can be mixed with other agents (see below). The hypnotic (sleep-inducing) effects of these agents in animals, unlike in humans, are generally minimal. When administered alone, benzodiazepines have a hyperalgesic effect in humans in some circumstances (i.e. they increase the degree of pain which is perceived). This may occur in animals and so they should not be used for post-operative sedation unless effective analgesia is also provided, e.g. by administration of opioids.

D. Alpha-2-adrenergic agonist tranquillizers: xylazine and medetomidine

1. Desirable effects

Xylazine and medetomidine are potent sedatives, and are hypnotics in some species. Their analgesic effects vary in different species, but in most animals mild to moderate analgesia is produced. Xylazine and medetomidine markedly potentiate the action of most anaesthetic drugs.

2. Undesirable effects

High doses of these drugs produce cardiovascular and respiratory depression and cardiac arrhythmias may occur following administration of xylazine in some species. Xylazine may cause severe respiratory depression if administered in combination with barbiturates or alphaxalone/alphadolone.

3. Special comments

Xylazine is a useful sedative in cattle, sheep, goats, horses, cats and primates. It may also be a valuable (but relatively short-acting) analgesic in sheep and goats. Both xylazine and medetomidine should be used with caution in sheep, since these drugs can produce severe hypoxia. The major use of xylazine in laboratory animal anaesthesia is in combination with ketamine to produce surgical anaesthesia (see Chapter 3). Xylazine has been reported to cause pronounced hyperglycaemia (Feldberg and Symonds, 1980) and a marked diuresis (Greene and Thurmon, 1988). A similar diuresis has been observed by the author during ketamine/xylazine anaesthesia in rats and mice. Medetomidine has similar effects to xylazine, but it is a much more specific alpha-2 agonist, and therefore is claimed to have a lower incidence of side-effects (Virtanen, 1989). It can be used to provide deep sedation with complete immobilization in many species, avoiding the need for general anaesthesia, and can be rapidly and completely reversed using the specific alpha-2 antagonist, atipamezole. Although reversal of xylazine with atipamezole is possible, since the antagonist has been developed to have a similar pharmacologic profile to medetomidine, it is preferable to use medetomidine if reversal is contemplated (see Chapter 3). Other related compounds, e.g. detomidine, are available for use in horses and ruminants, but there is little information available concerning their effects in small mammals.

III. NARCOTIC ANALGESICS

A. Morphine, pethidine, buprenorphine, butorphanol, nalbuphine, pentazocine, methadone, fentanyl, alfentanil, sufentanil, etorphine, oxymorphone

1. Desirable effects

These compounds can produce moderate sedation and profound analgesia but in some species pre-operative administration will cause hyperactivity and excitement. Further details of the effects of each agent are given in Chapter 6.

2. Undesirable effects

These drugs may produce respiratory depression (generally only at high dose rates) and vomiting in some species (dog, primates). A more detailed discussion of side-effects can be found in Chapter 6.

3. Special comments

Narcotic analgesics can be used both to provide pre-operative analgesia and to reduce the dose of anaesthetic agents necessary to produce surgical anaesthesia. It is also probable that pre-operative administration of analgesics may reduce the degree of post-operative pain (Woolf and Wall, 1986; McQuay *et al.*, 1988; Breivik, 1994). Opioids are also widely used as components of neuroleptanalgesic combinations (see Chapter 3). A more detailed description of the individual agents is given in Chapter 6.

3

Anaesthesia

To carry out surgical procedures on animals, pain perception must be completely suppressed. This can be achieved either by general anaesthesia, which produces loss of consciousness, or by local or regional anaesthesia. It is now recognized that whilst different agents can appear to provide similar levels of hypnosis (sleep), the degree of intraoperative analgesia provided can vary widely between agents. It is important that an anaesthetic regimen is selected that provides an appropriate degree of intra-operative analgesia. If anaesthesia is being induced simply to provide humane restraint while non-painful procedures are carried out, then only light anaesthesia, with little pain suppression, will be required. Conversely, if potentially painful surgical procedures are to be undertaken, then deep anaesthesia, with complete suppression of pain perception, is necessary.

I. GENERAL ANAESTHESIA

General anaesthesia can be induced using a variety of drugs and techniques. Often a single drug can be given to produce all the required features of general anaesthesia: loss of consciousness, analgesia, suppression of reflex activity, and muscle relaxation. Alternatively, a combination of agents can be given, each making a contribution to the overall effect. The advantage of such an approach is that the undesirable side-effects of anaesthetic agents can often be minimized. The side-effects of anaesthetics are usually dose dependent. Giving several drugs in combination, at relatively low dose rates, can often result in less effect on major body systems than that following induction of anaesthesia using a single anaesthetic agent. A brief review of some of the major effects of the more widely used anaesthetic agents is given below. More detailed reviews are available (Attia *et al.*, 1987; Short, 1987). Further discussion of factors influencing the selection of a particular method of anaesthesia is included in Chapter 4.

A. Administration of anaesthetic agents by inhalation

1. Equipment

a. Anaesthetic machines. Anaesthetic machines are designed to supply oxygen, anaesthetic gases and volatile anaesthetic agents to the animal. Even if volatile and gaseous anaesthetics are not used, it is desirable to supply oxygen to animals that have been anaesthetized with injectable agents. The design of different anaesthetic machines varies considerably, ranging from compact table-top models to large anaesthetic trolleys. These larger trolleys usually provide storage space for monitoring and other ancillary equipment, as well as a convenient table for items such as drugs and syringes. The gas supply may be provided from cylinders mounted on the anaesthetic trolley, or from a central gas supply by means of hose connections. Modern anaesthetic machines are fitted with a pin-index system of cylinder connectors to prevent inadvertent attachment of, for example, an oxygen cylinder to a nitrous oxide or other gas cylinder fitting. The gas cylinders or pressurized hose connectors deliver the gas via a pressure regulator to a flowmeter. Flowmeters enable control of the individual gas flows and allow the flow rates and relative proportions of nitrous oxide and oxygen to be controlled. The most common type of flowmeter consists of a bobbin mounted in a glass tube. As the flow increases, the bobbin rises higher up the tube. The gas flow can be read from the position of the top of the bobbin on a graduated scale on the tube. Some flowmeters are fitted with a second control which enables the percentage concentrations of oxygen and nitrous oxide to be pre-set.

The gases pass from the flowmeter through a vaporizer, which is filled with a volatile anaesthetic agent such as halothane. The vaporizer should deliver an accurate concentration of anaesthetic and this concentration should not be affected by changes in the gas flow rate or by changes in the temperature of the gases. Calibrated vaporizers of this type, such as the Tec models (Ohmeda, Appendix 7), are relatively expensive, but their use is essential if potent anaesthetics such as halothane or isoflurane are to be administered in a controlled way.

Calibrated vaporizers are designed for use with a particular volatile anaesthetic agent and it is important to ensure that only the appropriate agent is used. The latest designs of vaporizer incorporate a special filling device that both reduces the risk of spillage and pollution and prevents the attachment of bottles containing the incorrect anaesthetic agent.

Some older machines may be fitted with a simple Boyle's bottle vaporizer. This ducts gas over the surface of the liquid anaesthetic, so that some of the anaesthetic is vaporized. To increase the concentration of anaesthetic, the gases may be diverted to bubble through the liquid. Vaporizers of this type are only suitable for use with ether, methoxyflurane or trichloroethylene. They suffer from the serious disadvantage that changes in the gas flow rate or the temperature alter the concentration of anaesthetic that is delivered to the patient. Since the temperature of the liquid anaesthetic will fall as it is vaporized, there will be a marked tendency for the concentration of vapour produced to fall as the period of anaesthesia progresses.

b. Choice of apparatus. A fully equipped anaesthetic trolley is expensive and is unnecessarily sophisticated for some anaesthetic regimens. An additional problem is that the flowmeters provided may be incapable of delivering appropriately low flow rates. To overcome this problem, several designs for small rodent anaesthetic machines have been published (Sebesteny, 1971; Norris, 1981), and if the necessary technical expertise is available, these can be constructed at low cost. These simple machines generally make use of Boyle's bottle vaporizers and so are unsuitable for use with many newer volatile anaesthetics. To overcome this problem, a calibrated vaporizer should be purchased and economies made in the provision of the other components of the anaesthetic machine if necessary. Combined regulators and flowmeters are available from several companies (e.g. Ohmeda, Appendix 7), and designs are available that attach directly to pin-index cylinders. This avoids the necessity of maintaining stocks of the non-standard larger cylinders to which most regulators attach. The cylinder, with regulator and flowmeter attached, can be mounted on a suitable trolley (e.g. that available from I.M.S., Appendix 7) and connected to a vaporizer using antistatic hose and a catheter mount (Figure 3.1). An additional advantage of this system is that the trolley can be large enough to act as an operating table and can be equipped with a portable gas-scavenging system, homeothermic blanket, small rodent operating table and cold light source. The cost of such a well-equipped unit will usually be less than that of a standard commercially produced anaesthetic machine.

Whichever system is purchased or constructed, it must be carefully maintained. If a calibrated vaporizer is used it must be serviced regularly (approximately twice a year) to ensure that the concentrations of anaesthetic delivered correspond to the vaporizer settings.

c. Face-masks and endotracheal tubes. Whichever anaesthetic circuit is selected (see below), an endotracheal tube or face-mask will be required to connect it to the patient. The face-masks manufactured for veterinary use are cone-shaped and will be found adequate for sheep,

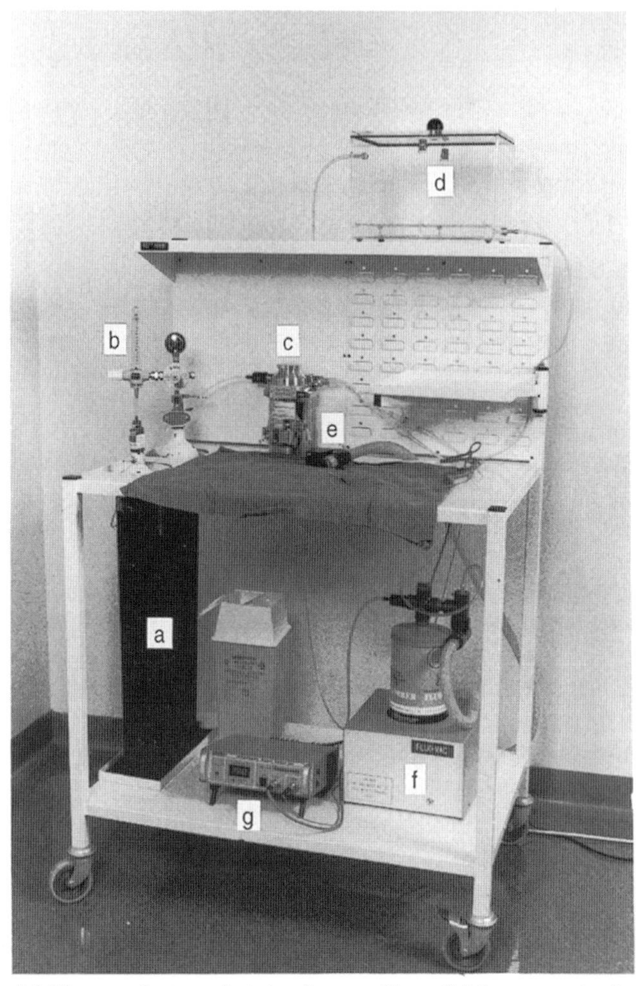

Fig. 3.1. Mobile anaesthetic workstation for use with small laboratory animals. Base unit available from IMS (Appendix 7). (a) Gas cylinder. (b) Combined pressure reducing valve and flow-meter. (c) Halothane vaporizer. (d) Anaesthetic chamber. (e) Face mask. (f) Scavenging system. (g) Heating blanket control unit.

pigs, dogs, cats and rabbits, provided that the appropriate size is used. Face-masks should fit snugly around the muzzle, must not obstruct the mouth or nose, but must provide minimal dead space. A suitable mask for rats and guinea pigs can be provided by using a human paediatric face-mask, the Charlotte's inhaler, available from M.I. & E. (Appendix 7), but

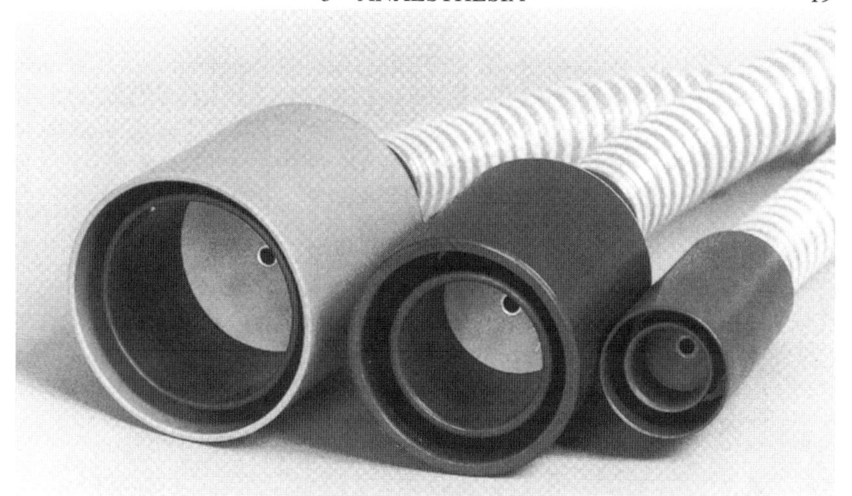

Fig. 3.2. Concentric mask system to enable effective gas scavenging. The three sizes of masks are suitable for mice, rats and rabbits, and animals of comparable size (available from IMS, Appendix 7).

mice and similar-sized animals will require purpose-made masks. A useful mask design that incorporates a gas-scavenging system has been described (Hunter *et al.*, 1984) and is available commercially (International Market Supply, Appendix 7) (Figure 3.2). An alternative system is also available (Viking Medical, Appendix 7). Since even the lowest flow rates that can be provided accurately by many anaesthetic machines are higher than those actually required by small rodents, most circuits that use a face-mask will be acting as open systems and the dead space in the face-mask will be relatively unimportant.

Face-masks should be cleaned after use by washing in warm, soapy water, followed by drying. They cannot be autoclaved but some may be sterilized using ethylene oxide. Endotracheal tubes are available from many manufacturers and are provided either as plain tubes, or with an inflatable cuff that seals the gap between the wall of the tube and the trachea (Figure 3.3). The cuff can be inflated either with a syringe (2.5 ml) or with a specially designed inflator. The cuff is prevented from deflating either by means of a non-return valve (present on some disposable tubes) or by clamping with a pair of haemostats. Tubes may be re-useable or be intended only for single use. Re-useable tubes are generally constructed of rubber and are opaque. They deteriorate gradually, becoming brittle and easily kinked. The cuff often becomes distorted and may leak. It is preferable to purchase single-use tubes and allow a limited amount of

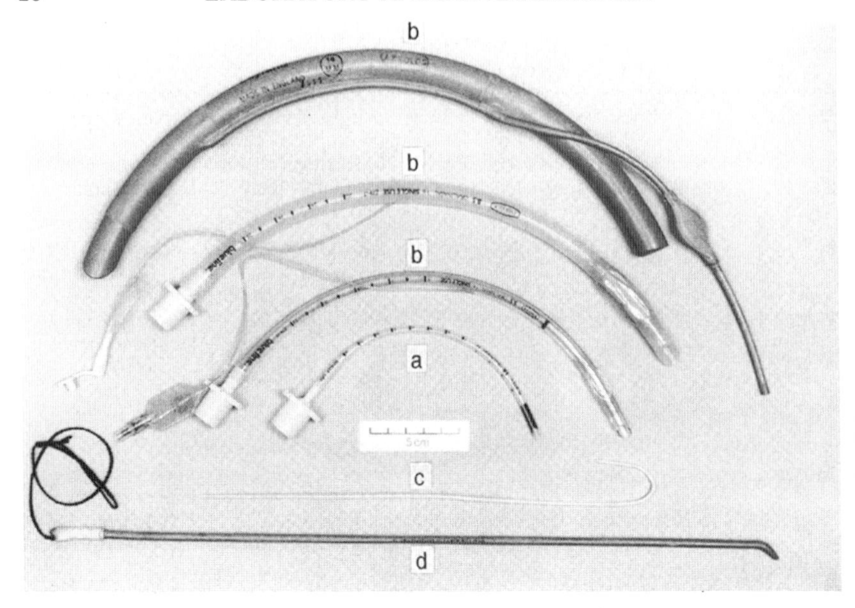

Fig. 3.3. Endotracheal tubes. (a) Uncuffed tube (3mm OD). (b) Cuffed tubes (5.5, 8.5 and 14mm OD). (c) Small (paediatric) introducer. (d) Adult size introducer.

re-use. Clear polyethylene tubes have the advantage that condensation appearing in the tube with each expiration provides an immediate indication that the tube is correctly positioned in the airway. Most commercially available tubes are too long for animal use and should be shortened as described below to reduce unnecessary dead space.

Tubes should be inspected carefully before use to ensure that they have not begun to deteriorate. They should be cleaned after use by washing in hot, soapy water, then thoroughly rinsed and dried. If apparatus for pasteurization is available, tubes can be pasteurized. Many types do not withstand autoclaving, although some may be autoclaved a limited number of times at lower temperatures (121°C for 15 minutes), or sterilized using ethylene oxide. Anaesthetic circuits and reservoir bags should be washed in hot, soapy water and either pasteurized or rinsed with chlorine disinfectant. Metal components can be autoclaved after washing.

d. Laryngoscopes. Laryngoscopes are used to obtain a clear view of the larynx so that an endotracheal tube may be passed easily and atraumatically. A variety of designs are available commercially, and a list of recommended blades is given in Table 3.1 and illustrated in Figure 3.4. The handle, besides usually containing the batteries, acts as a counter-

TABLE 3.1
Endotracheal intubation: equipment required

Species	Body weight	Endotracheal tube diameter	Laryngoscope
Cat	0·5–1·5 kg > 1·5 kg	2–3 mm O/D* 3–4·5 mm O/D	Macintosh size 1
Dog	0·5–5 kg > 5 kg	2–5 mm O/D 4–15 mm O/D	Macintosh size 1–4
Guinea pig	400–1000 g	16–12 gauge plastic cannula	Purpose-made laryngoscope† Otoscope
Hamster	120 g	1·5 mm	Purpose-made laryngoscope†
Mouse	25–35 g	1·0 mm	Purpose-made laryngoscope† Otoscope
Primate	< 0·5 kg 0·5–20 kg	Not reported 2–8 mm O/D	 Macintosh size 1–3
Pig	1–10 kg 10–200 kg	2–6 mm O/D 6–15 mm O/D	Soper or Wisconsin size 1–4
Rabbit	1–3 kg 3–7 kg	2–3 mm O/D 3–6 mm O/D	Winconsin size 0–1
Rat	200–400 g	18–12 gauge plastic cannula	Purpose-made laryngoscope† Otoscope
Sheep	10–90 kg	5–15 mm O/D	Macintosh size 2–4

* O/D, outside diameter
† Costa *et al.* (1986).

balance to the blade. For this reason it will be found most convenient to purchase handles of the appropriate size for each range of blade sizes. Replacement bulbs should also be purchased so that they are always readily available. Details of the use of a laryngoscope are given below. After use, the handle should be separated from the blade and wiped clean. The blade should be washed in hot, soapy water and dried thoroughly.

2. *Methods of administration of inhalational anaesthetics*

a. Anaesthetic chambers. It has been common practice to anaesthetize small rodents by placing them in a glass receptacle containing a pad of gauze or cotton wool soaked in liquid anaesthetic. Direct contact with the

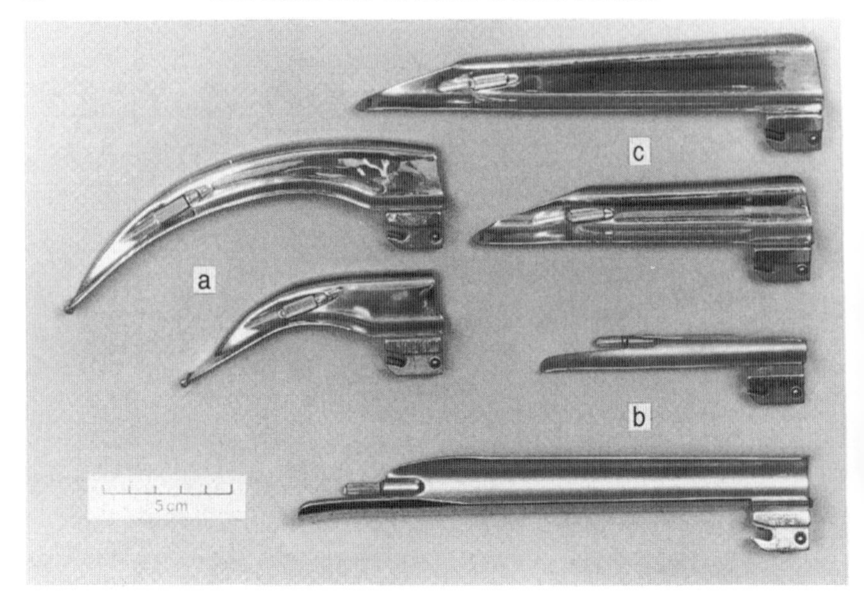

Fig. 3.4. Laryngoscope blades. (a) Macintosh size 2 and 4. (b) Wisconsin size 1 and 4. (c) Soper size 2 and 3.

liquid anaesthetic is extremely unpleasant for the animal, as it is irritant to mucous membranes. Even if the gauze is separated from the animal by a metal grid, liquid anaesthetic is often spilt on to areas that are in contact with the animal. The concentration of anaesthetic that can be achieved in such containers is unpredictable and is invariably dangerously high if potent, easily vaporized anaesthetics such as halothane are used. For example, the concentration of halothane produced at 20°C is 32%, more than six times the safe induction concentration (see Tables 4 and 5). If ether is used, there will be a significant risk of fire or explosion. Whichever volatile anaesthetic is used, it is frequently impossible to prevent contamination of the environment with anaesthetic vapour and this may present a hazard to the anaesthetist. Use of such an anachronistic technique has no advantage other than the low cost of the apparatus. A considerably safer and often more humane technique is to use an anaesthetic machine to deliver a known concentration of anaesthetic, using oxygen (either alone or in combination with nitrous oxide) as the carrier gas. A suitably sized (30 cm × 20 cm × 20 cm) Perspex box should be used, so that the animal can be observed during induction (Figure 3.5). The waste anaesthetic gas should be removed in a controlled way and either ducted out of the room or adsorbed using activated charcoal. A particularly effective scavenging technique has been devised using a double-box system (International Market Supply, Appendix 7) (Figure 3.6). A pad of towelling or Vetbed (Cox Surgical,

Fig. 3.5. Anaesthetic induction chamber for small animals (< 1kg). Note that the waste gas is removed from the top of the chamber and fresh anaesthetic gas introduced at the bottom.

Appendix 7), or some paper towels should be placed on the floor of the chamber and the apparatus should be cleaned thoroughly after use.

Since anaesthetic gases are denser than air, the anaesthetic chamber will fill gradually. To avoid gas-scavenging systems removing anaesthetic as rapidly as it is added, waste gas should be removed from the top of the chamber, and fresh gas should flow in at the base or the top.

To provide rapid induction, the entire chamber should be filled quickly. An appropriate gas flow rate can be estimated by measuring the chamber and calculating its volume. The time to fill the chamber completely can then be determined from the flow rate of anaesthetic:

Chamber volume/flow rate = time to fill

For example, a flow rate of 4 litres per minute, into a typical rodent induction chamber with a volume of 12 litres would give a filling time of 3 minutes. Filling times for larger chambers (for example those designed for use with rabbits and cats), with volumes exceeding 50 litres, can therefore be considerable.

b. Anaesthetic circuits. Anaesthetic chambers are useful for inducing anaesthesia in small animals which may be difficult to restrain, but the animal must be removed from the chamber to enable surgical manipulations

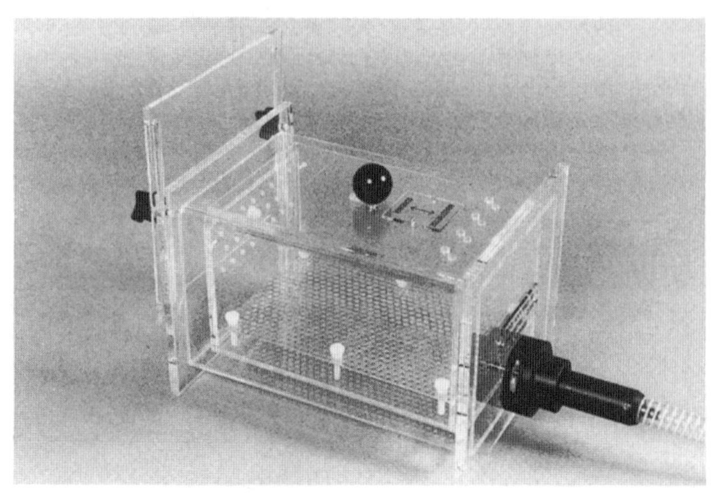

Fig. 3.6. Double chamber system for anaesthetizing small mammals. All anaesthetic gases can be removed from the system before opening the chamber. (IMS Ltd., Appendix 7).

to be carried out. Unless the operative procedure is of extremely short duration, some method of maintaining anaesthesia must be provided.

The first impression of anaesthetic breathing circuits is that they are complex and difficult to use. This apparent complexity, coupled with a reluctance to develop the expertise necessary to carry out endotracheal intubation, has led to an over-reliance on open-mask systems. When other circuits are used, they may be incorrectly assembled or may be inappropriate for use with small animals. Several advances in circuit design have been introduced into human anaesthetic practice and these developments can often usefully be transferred to laboratory animal anaesthesia.

i. General considerations. All anaesthetic breathing systems aim to deliver sufficient anaesthetic gases to meet the animal's ventilatory requirements and to remove exhaled gases which will contain carbon dioxide. It is an advantage if these exhaled gases can be removed from the operating area, as trace concentrations of anaesthetic gases may have adverse effects on operating theatre personnel. An additional consideration when selecting a breathing system is the ease and efficiency with which assisted ventilation can be carried out.

Two concepts discussed below are those of *dead space* and *circuit resistance.* The dead space of an anaesthetic circuit is the part of the circuit that remains filled with expired gas at the end of expiration, and this carbon dioxide-rich gas is then re-inhaled by the animal. If a significant amount of expired gas is rebreathed the blood carbon dioxide concentration will rise and produce a range of adverse effects (see Chapter 4). The resistance of an anaesthetic circuit influences the effort that must be made by the animal to move gas in and out of the circuit. Circuits with narrow or sharply angled components and those with valves will provide a greater resistance to gas flow and so will require a greater respiratory effort by the animal.

It is also necessary to use the terms *tidal volume* and *minute volume*. The tidal volume is the volume of gas drawn into the respiratory tract with each breath. The minute volume is the volume of gas drawn into the respiratory tract in 1 minute and so is calculated by multiplying the tidal volume by the respiratory rate. The minute volume is not always equivalent to the flow rate of gas that needs to be delivered to the animal by the anaesthetic circuit. During each respiratory cycle, gas is only drawn into the lungs for approximately one-third of the time during inspiration and obviously not during expiration or during the pause between expiration and the next inspiration. This means that the animal's minute volume is inhaled in approximately 20 seconds, so using a simple face-mask system, a fresh gas flow rate of three times this volume per minute is required. Occasionally an even greater flow rate, to meet the most rapid rate at which gas is drawn into the lungs, the peak inspiratory flow rate, may be needed. Many anaesthetic circuits are designed to economize on fresh gas flow rates by providing a reservoir for the unused gas delivered by the anaesthetic machine.

ii. Open breathing circuits. As mentioned earlier, the most widely used breathing circuit is an open face-mask (Figure 3.7), which is a simple and convenient way of delivering anaesthetic gases to an animal. Expired gases pass round the edges of the masks and, provided that the gas flow rate is sufficiently high, rebreathing of exhaled gas will be small and dilution of the anaesthetic gases by breathing room air will be avoided. If the gas flow

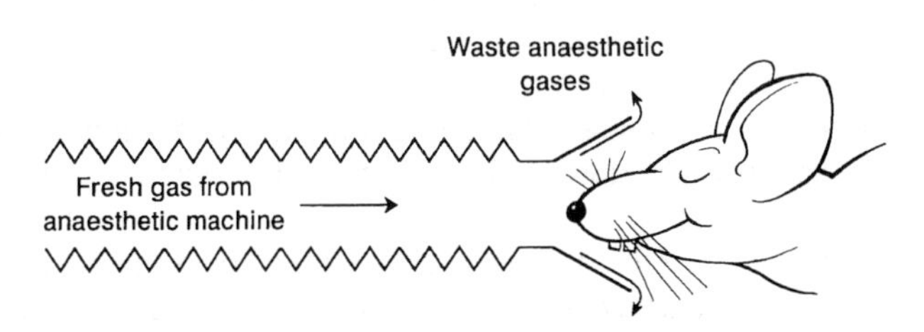

Fig. 3.7. Open anaesthetic circuit with delivery of anaesthetic gases by means of a face-mask. Expired gases pass around the mask.

TABLE 3.2
Recommended fresh gas flow rates for different anaesthetic circuits

Body weight (kg)	Estimated tidal volume (ml)	Minute volume (l)	Flow rate, open system ($l\ min^{-1}$)	Flow rate, T-piece or Bain circuit ($l\ min^{-1}$)	Flow rate, Magill circuit ($l\ min^{-1}$)	Flow rate closed circuit* l/min
0·5	5–7·5	0·4–0·6	1–2	1–1·5	–	–
1	10–15	0·5–1	1·5–3	1·5–2·5	–	–
3	30–45	1–1·5	3–4·5	2·5–3·5	–	–
6	60–90	1·5–3	4·5–9	3·5–7·5	–	–
10	100–150	3–6	9–18	6–12	3–6	0·3
20	200–300	5–9	15–27	10–18	5–9	0·5

* See text for additional information.

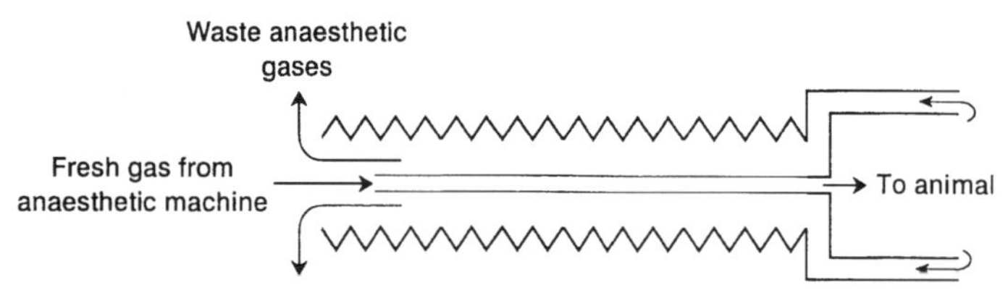

Fig. 3.8. Waste anaesthetic gas-scavenging system for use with small laboratory animals.

rate is too low, the animal will breathe in room air from around the edges of the mask. This will result in a reduction in the depth of anaesthesia if volatile anaesthetics are being used. As mentioned above, to meet inspiratory flow requirements, the gas flow rates must be three times the animal's minute volume. Typical flow rates are shown in Table 3.2. Use of this simple method of delivery has several drawbacks. The gas flow rate must be high, although when animals weighing less than 1 kg are anaesthetized, this is relatively unimportant, particularly since many older anaesthetic vaporizers are inaccurate at flow rates of less than 1 litre per minute. Removal of waste anaesthetic gases is difficult because gas escapes all around the mask. Scavenging hoods have been devised, but these may interfere with the surgeon's view of the animal. The most satisfactory technique is to use the scavenging system described by Hunter *et al.* (1984) (Figures 3.1, 3.2 and 3.8), available from International Market Supply (Appendix 7). Although this might appear to resemble the coaxial Bain circuit described below, it must be stressed that the outer tube does not act as a gas reservoir and so gas flow rates appropriate to an open system must be used (Table 3.2). Perhaps the most serious disadvantage in using a simple open circuit is that it is not possible to assist ventilation artifically should this be required.

iii. Semi-closed breathing circuits. Semi-closed breathing circuits are systems in which some rebreathing of expired gases may occur and in which no carbon dioxide absorption is used.

iv. The T-piece circuit. The T-piece circuit was first described by Ayre (1956) and was intended to provide a circuit with low resistance and little dead space, for use in infants and young children. The circuit consists of a tube into which the anaesthetic gas mixture is introduced through a small inlet tube at right angles to the main limb (Figures 3.9 and 3.10). One end of the T-piece is connected to the animal, while the other is left open to the air. A length of tubing is attached to this open end, providing a small reservoir for anaesthetic gases which would otherwise escape into the outside air. The presence of this reservoir enables the fresh gas flow to be reduced to about twice the animal's minute volume, without rebreathing occurring (see Table 3.2). During inspiration, fresh gas is drawn in both from the side-arm and from the reservoir. During expiration, exhaled gas fills the reservoir limb and during the phase before the next inspiration, is washed out by the fresh gas from the side-arm. The volume of the reservoir limb is unimportant, so long as it exceeds one-third of the animal's tidal volume and does not impose any appreciable resistance to expiration.

Ventilation can be controlled simply by intermittently occluding the end of the reservoir limb, but if this is done manually the anaesthetist has very little idea of the pressure being delivered to the animal's lungs. It is preferable to attach an open-ended reservoir bag to the expiratory limb: the Jackson–Rees modification (Rees, 1950). Squeezing the bag with the end occluded inflates the lungs and exhalation occurs through the open end of the bag (Figure 3.9). No increase in fresh gas flow rate is required when assisting ventilation in this way. It is easy to attach a mechanical ventilator to the reservoir limb and ventilate the limb with air. Provided that the reservoir is of sufficient volume, little or no mixing of the anaesthetic gases and the ventilating gas occurs.

To use the T-piece effectively it should be connected directly to an endotracheal tube or to a close-fitting face-mask. The volume of the patient side of the T should be low to reduce equipment dead space. Similarly, the volume of endotracheal tube connectors should be minimized. This is best achieved by using Oxford paediatric connectors and a Bethune T-piece (Penlon Ltd, Appendix 7) or other, disposable, paediatric connectors, such as the Minilink system (Portex Ltd, Appendix 7). These

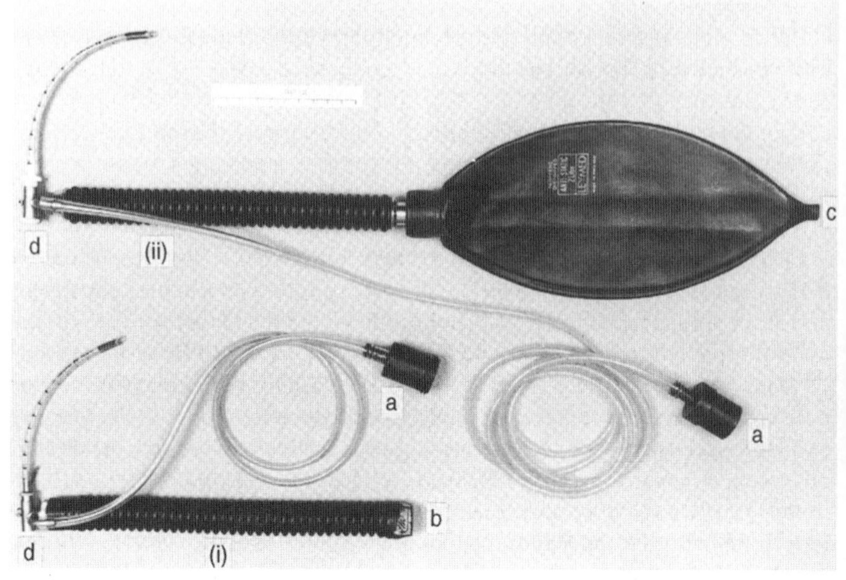

Fig. 3.9. (i) Ayre's T-piece. The circuit should be connected to the animal by means of an endotracheal tube or close-fitting face-mask. (ii) Jackson-Rees modified T-piece. (a) Fresh gas supply. Waste anaesthetic gases can be scavenged from the open end of the reservoir tube (b) or the open end of the reservoir bag (c). Note the use of low dead space endotracheal tube connectors (d).

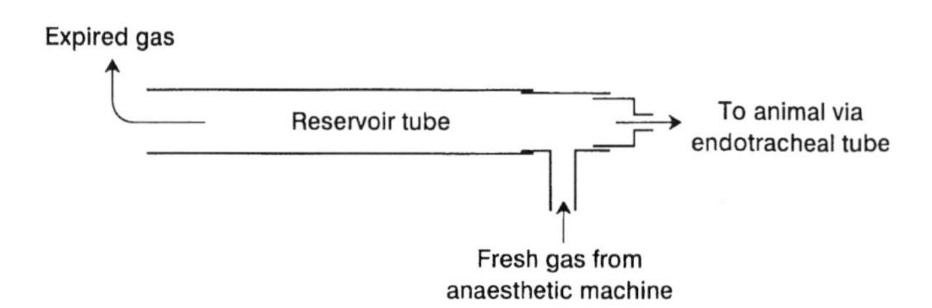

Fig. 3.10. Ayre's T-piece.

connectors contribute a dead space of approximately 0.2 ml compared with 1.5 ml when using a conventional type of connector (see Figures 3.3, 3.9 and 3.19). The dead space of a Bethune T-piece is only 1.1 ml. If a face-mask is used, it is essential that it fits closely around the animal's muzzle. If it does not, gas will be drawn in around the edges of the mask and the anaesthetic gas mixture will be diluted with room air. If ventilation is assisted, gas will escape around the mask and the degree of lung inflation produced will be inadequate. For these reasons it is preferable to intubate the animal whenever possible.

The T-piece is an ideal circuit for small laboratory animals since it offers low resistance to breathing and has a small dead space. It is not always necessary to purchase commercially produced T-pieces as the apparatus can be constructed easily from plastic 'T' connectors (Portex Ltd, Appendix 7) and rubber tubing.

v. The Bain coaxial circuit. The Bain circuit is a coaxial version of a T-piece, in which the fresh gas inflow tubing runs inside the reservoir limb (Figures 3.11 and 3.12). The circuit was designed to provide a lightweight circuit in which any valves or breathing bags were situated some distance from the patient and close to the anaesthetic machine (Bain and Spoerel, 1972). The lightweight contruction reduces the tendency for the circuit to pull on the endotracheal tube and so reduces the risk of inadvertent extubation. Positioning the expiratory port well away from the patient allows ventilation to be assisted easily without interfering with sterile drapes or the activities of the surgeon. In addition, anaesthetic gases can be scavenged easily and do not accumulate at the surgical site. The circuit has a small dead space (< 2 ml) and so is suitable for use in small animals. The Bain circuit functions similarly to a T-piece. During inspiration gas is drawn in from the central fresh gas supply and also from the outer reservoir tube. During expiration, exhaled gas fills the reservoir tube and during the pause before the next inspiration, this is replaced with fresh gas, provided that the fresh gas flow rate is adequate.

Two modifications of the basic circuit have been described. The expiratory limb may terminate with a pop-off valve and a reservoir bag (Figures 3.11 and 3.12b), or an open-ended reservoir bag may be mounted at the end of the expiratory limb. The former modification is unsuitable for small animals (< 10 kg body weight), since the presence of the valve increases circuit resistance. When used with larger animals, the valve and reservoir bag allow ventilation to be assisted easily by partially closing the valve and intermittently squeezing the reservoir bag. The use of an open-ended reservoir bag is equivalent to the Jackson–Rees modified T-piece and serves a similar function in allowing easy control of ventilation. Mechanical ventilators can be connected to the reservoir limb, as with a T-piece and the limb ventilated with air.

The gas flow rates required to prevent rebreathing have been quoted as ranging from 100 ml per kg of body weight per minute (Manley and McDonell, 1979a) to 200–300 ml kg^{-1} min^{-1} (Ungerer, 1978). The apparent adequacy of the lower flow rates can be explained by the animal responding to changes in its blood carbon dioxide concentration by altering its rate and depth of respiration. When the fresh gas flow is low, some rebreathing of exhaled carbon dioxide from the reservoir limb will occur.

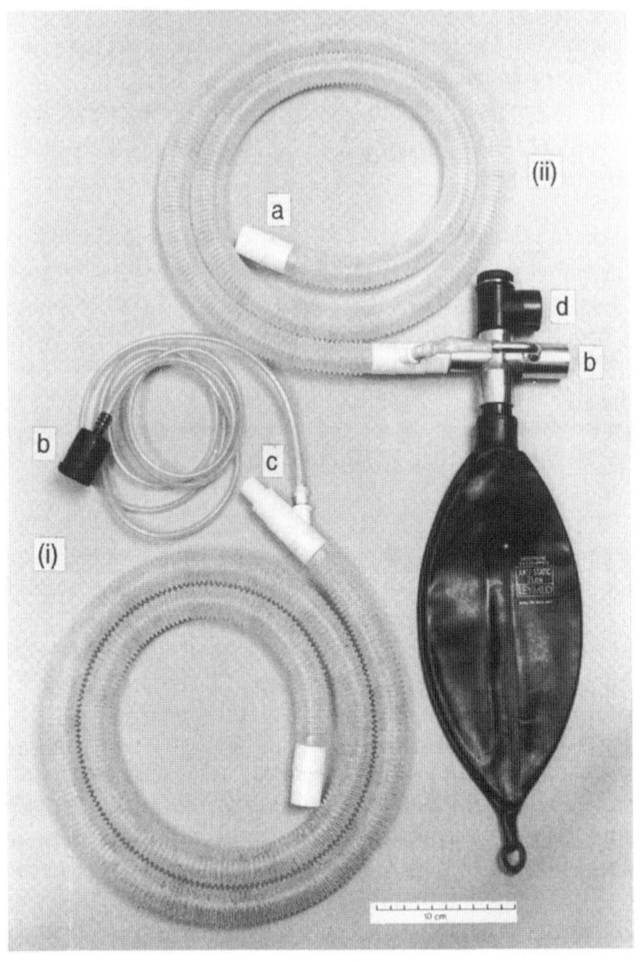

Fig. 3.11. (i) Bain (coaxial) circuit and (ii) Modified Bain (coaxial) circuit. The circuit should be connected to the animal by means of an endotracheal tube or close-fitting face mask at (a). (b) Fresh gas supply. Waste anaesthetic gases can be scavenged from the open end of the resevoir tube (c) or from the expiratory valve (d).

This will result in an increase in the blood carbon dioxide concentration which stimulates respiration. This moderate hyperventilation results in blood carbon dioxide tensions being maintained at acceptable levels. It is not certain whether the additional respiratory effort this causes is deleterious to the patient but the conventional view has always been that rebreathing should be minimal during spontaneous respiration. For this reason, it is recommended that fresh gas flow rates of 2–2.5 times the minute volume should be used (see Table 3.2). During mechanical ventilation some degree of rebreathing of carbon dioxide is advantageous, since it helps to avoid the

(a)

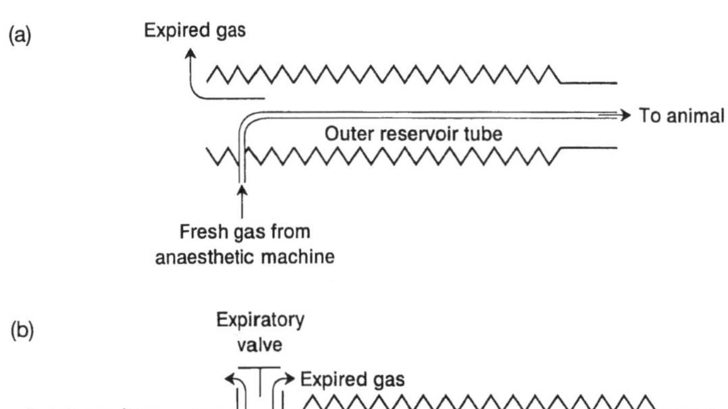

(b)

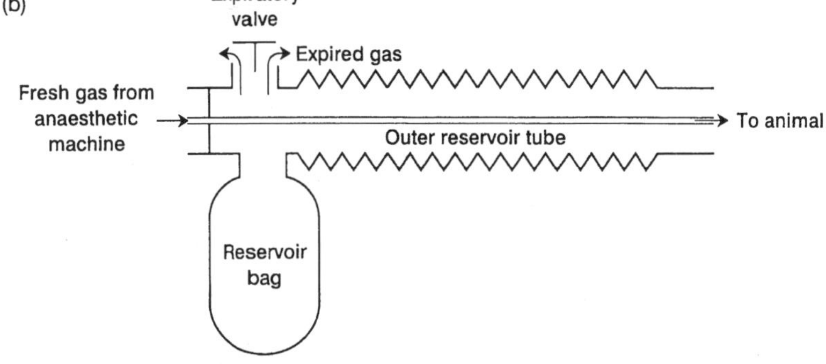

Fig. 3.12. (a) Bain (coaxial) circuit. (b) Modified Bain (coaxial) circuit.

production of hypocapnia. Fresh gas flow rates of 70–100 ml kg^{-1} min^{-1} allow the maintenance of normal carbon dioxide concentrations (normocapnia) during mechanical ventilation (Manley and McDonell, 1979b), but inaccuracies in rotameter settings limit the usefulness of this technique in small animals (Hird, 1977).

vi. The Magill circuit. The Magill breathing circuit is perhaps the most widely used circuit in human anaesthesia, so this probably accounts for the frequency with which it is used in animal anaesthesia. Whilst the advantages that have assured its popularity in human anaesthesia are applicable to similar-sized animals, it is generally unsuitable for use in animals weighing less then 10 kg.

The circuit consists of a reservoir bag connected by a length of corrugated tubing to the animal (Figures 3.13 and 3.14). An expiratory pop-off valve is situated as close to the patient as possible, to reduce equipment dead space. During expiration, the first portion of expired gas is from the animal's anatomical dead space (the trachea and bronchi) and, since no gas exchange occurs in this region, it contains no carbon dioxide. The expired gas travels up the corrugated tubing towards the reservoir bag which fills and then, as the pressure in the circuit rises, the expiratory valve lifts and

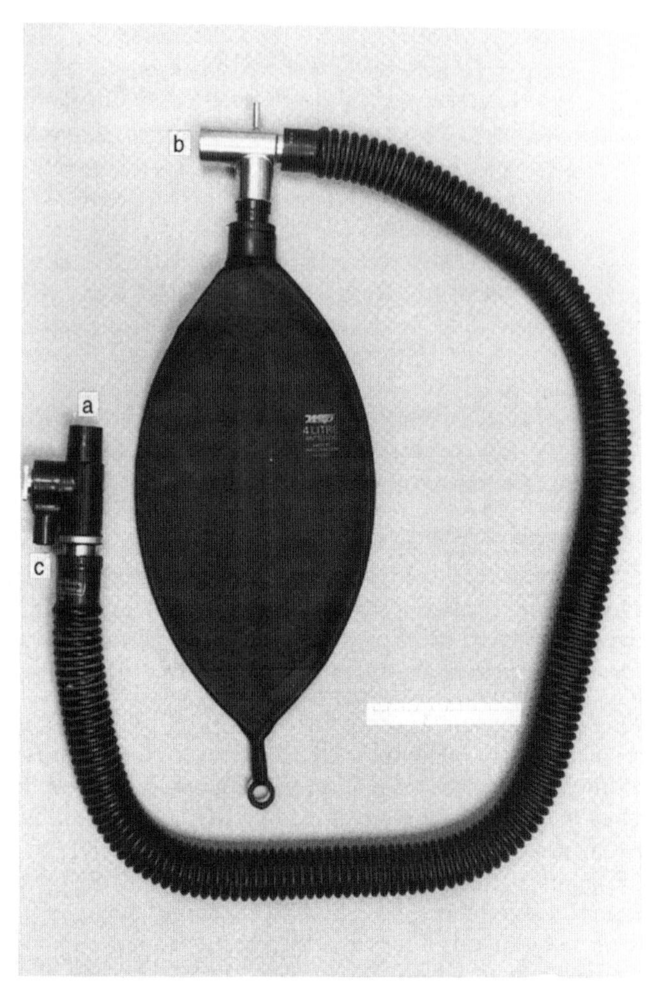

Fig. 3.13. Magill circuit. The circuit should be connected to the animal by means of an endotracheal tube or close-fitting face-mask at (a). Fresh gas supply enters at (b). Waste anaesthetic gases can be scavenged from the expiratory valve (c).

the remaining expired gas passes out of the circuit. The continued flow of fresh gas down the circuit flushes out any remaining carbon dioxide-rich alveolar gas during the pause before the next inspiration. Because of the preferential elimination of carbon dioxide-rich alveolar gas, significant rebreathing does not occur in humans until the fresh gas flow falls below 70% of minute volume (Kain and Nunn, 1968); the circuit is therefore extremely economical in its fresh gas requirements. It is important to realize that during controlled ventilation achieved by manual compression

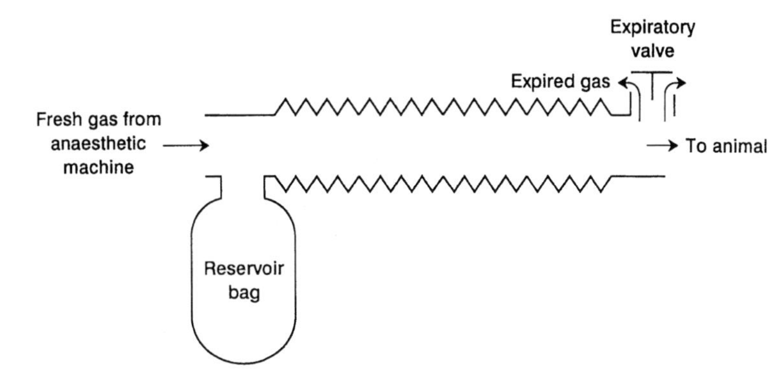

Fig. 3.14. Magill circuit.

of the reservoir bag this preferential elimination of alveolar gas is lost; under these conditions fresh gas flows of three times minute volume may be required to prevent rebreathing. The major problem in using the circuit with small animals is that it imposes a significant resistance to expiration. In addition, the dead space of a typical circuit is 8–10 ml, which is likely to represent a significant proportion of the tidal volume of a small animal. If it is to function effectively, the Magill circuit must be attached to the animal by an endotracheal tube or a close-fitting face-mask. It is common practice to connect this system to small animals with a badly fitting mask. Under these circumstances the circuit functions as an open system so that fresh gas flows in excess of three times minute volume are required to prevent either rebreathing or the dilution of the inspired gas mixture by room air, which will be drawn in around the face mask. When used correctly, waste anaesthetic gases can be scavenged by means of a suitable attachment on the expiratory valve. If used with a badly fitting face-mask, the same problems of pollution arise as occur with open circuits.

vii. Closed breathing circuits. Closed breathing circuits are systems in which the expired carbon dioxide is absorbed, usually by means of a soda lime canister. Because of the considerably lower fresh gas flows required, closed circuits are often used when anaesthetizing larger animals (body weight > 20–30 kg). The use of such circuits poses considerable problems for the inexperienced anaesthetist, and expert advice and assistance should be obtained before attempting to employ these techniques. Two types of circuit are available, the 'circle' system (Figures 3.15 and 3.16) and the 'to and fro' system (Figures 3.17 and 3.18). The latter design has the carbon dioxide absorber placed between the animal and a reservoir

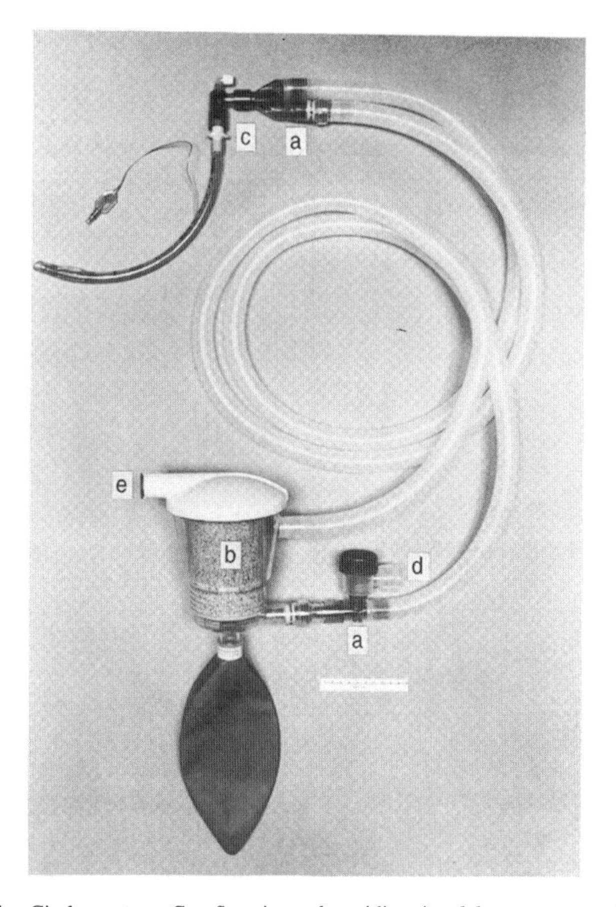

Fig. 3.15. Circle system. Gas flow is made unidirectional by one way valves at (a). Carbon dioxide is removed by the soda lime cannister (b). The circuit should be connected to the animal by means of an endotracheal tube or close-fitting face-mask (c). Surplus anaesthetic gases can be scavenged from the pop-off valve (d). The disposable circuit illustrated uses an out-of-circle calibrated vaporizer to supply fresh anaesthetic gases at (e).

bag. Respiratory movements result in gas being breathed out over the soda lime and into the reservoir bag, then back over the soda lime and into the animal's lungs. As anaesthesia progresses, the soda lime closest to the animal becomes exhausted, effectively increasing the dead space of the circuit. To reduce this problem, the canister may be turned around half-way through the period of anaesthesia. To and fro systems have decreased in popularity and greater use is now made of circle systems. In a circle system, the flow of gas is controlled by two unidirectional valves. These

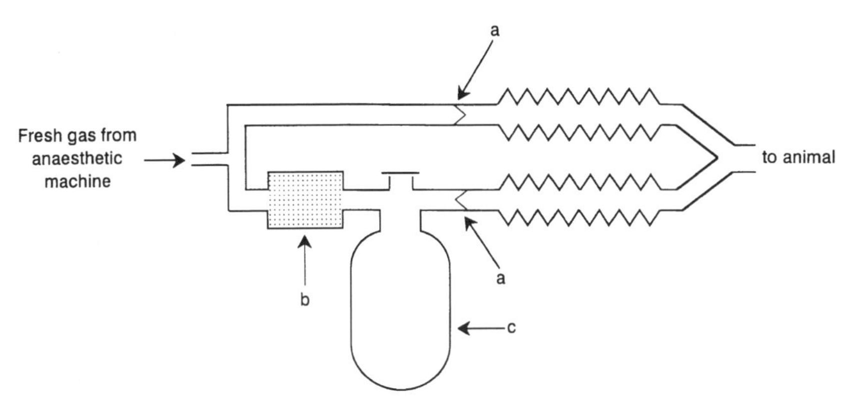

Fig. 3.16. Circle system. Gas flow is made unidirectional by one way valves at (a). Carbon dioxide is removed by the soda lime cannister (b). Reservoir bag (c) is situated below the pop-off valve.

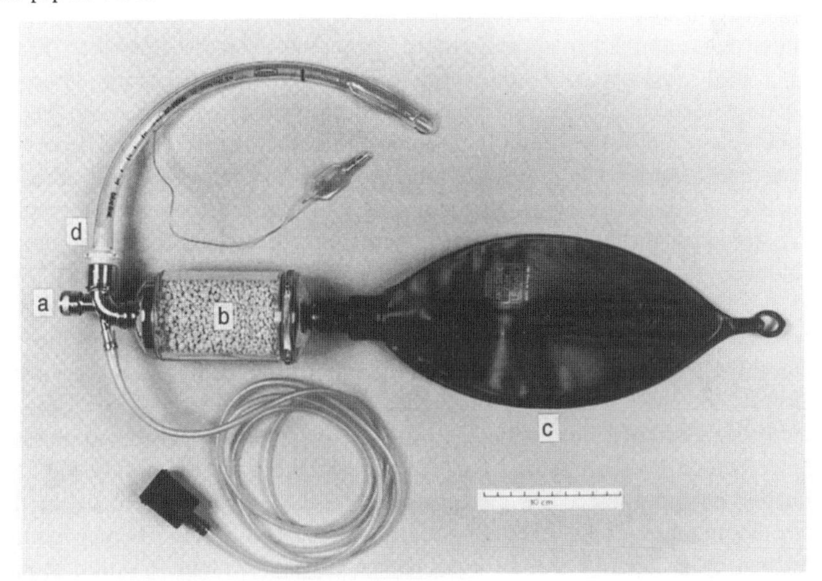

Fig. 3.17. To and fro system. Surplus anaesthetic gases can be scavenged from the pop-off valve (a). Carbon dioxide is removed by the soda lime cannister (b) as it passes to and from the reservoir bag, (c). The circuit should be connected to the animal by means of an endotracheal tube or close-fitting face-mask (d). Fresh anaesthetic gases enter via the tubing below the pop-off valve (a).

cause expired gas to pass through a soda lime canister, where carbon dioxide is absorbed, before the gases pass around the circuit to be breathed in again by the animal.

When using a closed circuit, it would be possible to supply only the

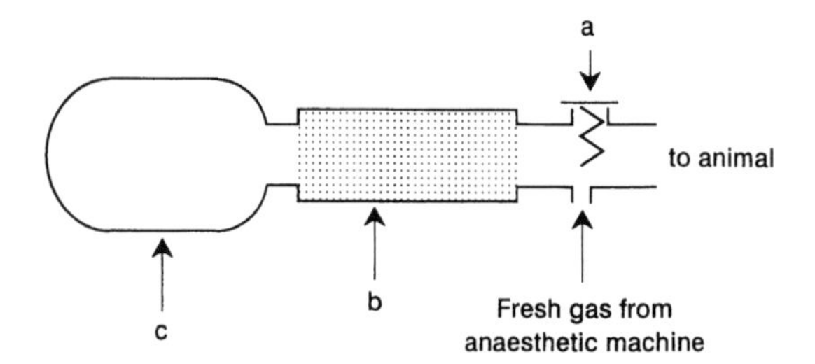

Fig. 3.18. To and fro system. Surplus anaesthetic gases can be scavenged from the pop-off valve (a). Carbon dioxide is removed by the soda lime cannister (b) as it passes to and from the reservoir bag (c).

animal's metabolic oxygen requirements (approximately 10 × body weight $(kg)^{0.75}$, i.e. 6–9 ml kg^{-1} min^{-1} for a 3 kg animal) (Brody, 1945), but for practical reasons it is more usual to operate the circle system as a low-flow rather than completely closed circuit. Typically, fresh gas flows of 100 ml kg^{-1} min^{-1} are used for small animals (< 10 kg) and 20–30 ml kg^{-1} min^{-1} for larger animals. This represents a major advantage of circle systems in comparison with other circuits in controlling anaesthetic costs. The newer anaesthetic agents such as isoflurane are expensive, and in larger animals the use of a closed system, where appropriate, can reduce costs (Appendix 2). Although circle systems have been more frequently used for larger animals, the introduction of lightweight disposable systems with low-resistance valves has increased their use for smaller animals. Despite advances in circuit construction, the system still offers more resistance to breathing than a T-piece or Bain circuit, and it is advisable not to use these circuits on small animals (< 5 kg) unless mechanical ventilation is used. As mentioned earlier, inexperienced users are strongly advised to seek assistance before using closed circuits. Two important points should be noted. If nitrous oxide is used, the concentration of this gas can build up in the circuit, resulting in a dangerously low concentration of oxygen. Either nitrous oxide should not be used, or an oxygen content monitor should be included in the circuit. When using a closed circuit, the concentration of volatile anaesthetic in the circuit will not be the same as that shown by the vaporizer setting. This can result in a failure to maintain adequate depths of anaesthesia. As experience is gained, the vaporizer setting can be increased to compensate for dilution of anaesthetic in the circuit, and uptake by the animal. A more reliable technique is to purchase an anaesthetic gas analyser. The cost of these monitors has fallen considerably, and a calculation of the savings in volatile agent that would result may show that they represent a worthwhile investment.

Besides economic considerations, another advantage of closed or low-flow circuits is that heat and moisture are conserved. A detailed comparison of the advantages and disadvantages of rebreathing and non-rebreathing circuits has been given by Brouwer and Snowdon (1987).

The diagram of the circle system in Figure 3.16 has the vaporizer placed outside the main circuit. In-circle vaporizers can also be used, but a full discussion of the relative merits of each arrangement is outside the scope of this book. A full description of these circuits is given by Steffey (1994) and Hall and Clark (1991).

viii. Recommendations. Although open mask techniques are best used only for short procedures in large animals, they may often be the most convenient system for small rodents, when the higher fresh gas flow rates required by these circuits ($< 1.5\ \mathrm{l\ min^{-1}}$) will be of little significance. If nitrous oxide or volatile anaesthetic agents are used, the provision of effective gas scavenging must be considered essential. For larger animals, such as the cat, rabbit and non-human primate, the advantage of lower fresh gas flow requirements, ease of gas scavenging and ability to assist ventilation favour the selection of a more sophisticated breathing circuit. If an Ayre's T-piece is used, it is strongly recommended either that a human paediatric model with a low dead space is obtained, or that one with similar features is constructed. The Bain circuit offers several advantages for use in small animals, particularly in respect of its low weight and small dead space. The ease with which controlled ventilation can be carried out from a point remote from the surgical field is a further distinct advantage. In addition, it is suitable for use both with small animals such as guinea pigs and rabbits and with larger species such as dogs. The Magill circuit is not suitable for use in animals weighing less than 10 kg, but it may be used as an alternative to the Bain circuit in larger animals such as dogs, sheep and pigs. Prolonged anaesthesia of larger animals (> 10 kg), particularly when using expensive agents such as isoflurane, is best provided using a circle system. Rather than attempt to use a fully closed system, a moderate fresh gas flow of $500\text{–}1000\ \mathrm{ml\ min^{-1}}$ should be used, as this will simplify use of the circuit (see above).

All the commonly used anaesthetic circuits are now available as lightweight, single-use items for human anaesthesia (Appendix 7). Many of these disposable circuits can be re-used on numerous occasions without difficulty, but it is essential that a careful check is made of the condition of the circuit each time it is used. In particular, ensure that any pressure relief valves are functioning correctly, and when using a Bain circuit check that the inner, fresh gas tube has not become disconnected at the anaesthetic machine end of the circuit.

c. Endotracheal intubation. Endotracheal intubation of large animals such as dogs, sheep, pigs, Old World primates and large birds (> 1 kg) is relatively straightforward, provided a suitable size and shape of laryngoscope is available (see Figure 3.4). A range of MacIntosh or Soper laryngoscope blades can be used for cats (size 1) and dogs (sizes 1–4), and a MacIntosh blade (sizes 2–4) for sheep. When anaesthetizing pigs, Soper (sizes 1–3) or Wisconsin (sizes 1–4) blades are preferable, although large pigs may require the use of a purpose-made laryngoscope blade. Rabbits can be successfully intubated using a Wisconsin blade (size 1 or 0) (see Table 3.1).

If intubation of a particular species is planned, a careful examination of the pharynx and larynx should first be carried out on a post-mortem specimen. This will enable an appreciation of the anatomical relationships within this area, particularly that of the soft palate and epiglottis. Once the normal anatomy of the region has been reviewed, a suitably sized endotracheal tube should be prepared. Most commercially available tubes are too long; their length should be reduced so that it approximates to the distance from the external nares to just anterior to the thoracic inlet. If a small (< 4 mm outside diameter) tube is to be used, an uncuffed tube is preferable as this enables the largest possible diameter tube to be passed. It is advisable to lubricate the tube with a small quantity of lignocaine gel.

The animal should be anaesthetized to a sufficient depth to abolish the cough and swallowing reflexes. It is possible to intubate lightly anaesthetized animals but, whilst this may be desirable under some circumstances, it is advisable to gain some proficiency in the technique of intubation before attempting this.

i. Dog, cat and sheep. The animal is placed in sternal recumbency and its jaws opened as widely as possible by an assistant. The tongue is drawn forwards and the laryngoscope advanced over the tongue towards the pharynx. The larynx is usually masked by the epiglottis. Gentle upwards pressure on the soft palate with the end of the endotracheal tube will disengage the epiglottis, allowing it to fall forwards, so providing an unobstructed view of the larynx. In the cat and the sheep the larynx should be sprayed with local anaesthetic, to prevent laryngospasm. In the UK, the most recent formulation of lignocaine (Astra) has been associated with the occurrence of laryngeal oedema (Taylor, 1992) and it is advisable to check the suitability of any locally available product before use. An alternative is to use lignocaine (2%) without adrenaline (0.25–0.5 ml per cat, equivalent to 5–10 mg total dose), delivered using a small atomizer or syringe and fine gauge needle. Disposable 'insulin' syringes with a pre-attached 25 G needle (see Figure 3.23), with the needle bevel cut off, are ideal. The endotracheal tube can then be advanced through the larynx and into the trachea. The tube is connected to the anaesthetic circuit, the cuff (if present) inflated and the tube tied in place to the animal's jaw using 1 cm wide cotton tape. It is preferable at this stage to assist ventilation (as described earlier) and observe that movement of both sides of the thorax occurs. This ensures that the tube has not been inadvertently positioned in one of the two bronchi. In addition, manual inflation of the chest will enable an appreciation of the degree of resistance to gas flow. Increased resistance may indicate any twisting or kinking of the tube, or its partial obstruction due to positioning close to the bifurcation of the trachea. If any uncertainty exists about tube placement, the chest should be auscultated using a stethoscope to check that breath sounds can be heard on both sides of the thorax.

ii. Pig. Intubation in the pig is complicated by the difficulty of obtaining an unobstructed view of the larynx. The animal is best positioned on its back, enabling the head and neck to be fully extended. As with other species, care must be taken when the tongue is extended to avoid damaging its surface on the teeth, particularly the canines in boars. Intubation is easier if an introducer is used (see Figure 3.3). This is a blunt stylet which is placed inside the tube to straighten it and make it easier to direct into the larynx. Introducers should be purchased commercially (Portex Ltd; M & IE Dentsply; Appendix 7), as it is important that the tip is soft and atraumatic. In an emergency, it is possible to construct an introducer from a suitable length of wire, but this increases the risk of laryngeal trauma.

The laryngoscope is advanced over the tongue and the epiglottis disengaged from the soft palate if necessary, by pushing downwards on the soft palate using the tip of the introducer. Once the larynx has been located it should be sprayed with lignocaine. The introducer and endotracheal tube can then be gently advanced into the larnyx and the introducer withdrawn. The tube should then be gently advanced; at this stage its progress is usually arrested by the laryngeal wall. If this occurs, the tube should be withdrawn very slightly, rotated through 90 degrees and re-advanced. This should be repeated as necessary until no resistance is experienced. Under no circumstances should attempts be made to pass the tube forcibly through the larynx as this is likely to result in severe trauma, haemorrhage and consequent asphyxiation.

iii. Rabbit. Visualization of the larynx in the rabbit is difficult, and it is necessary to use either a Wisconsin laryngoscope blade (size 1 in rabbits over 3 kg, size 0 in smaller rabbits) or an otoscope if intubation is to be

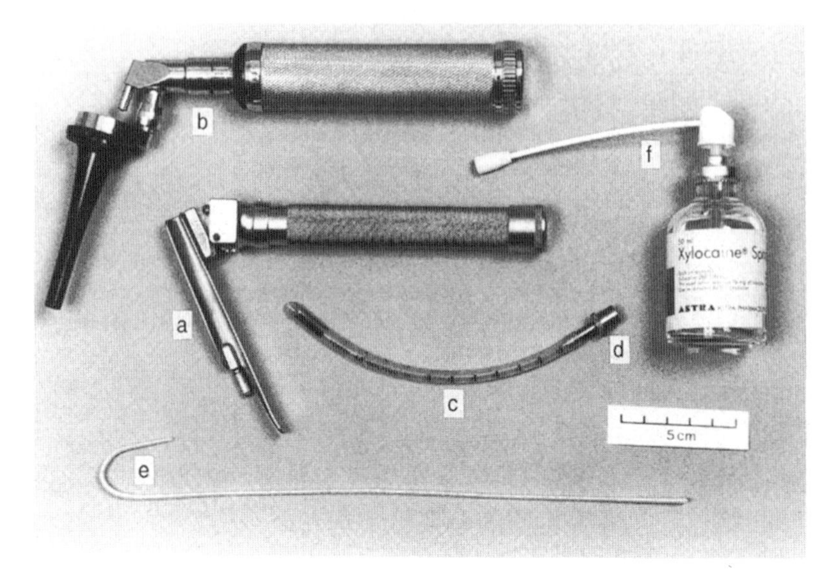

Fig. 3.19. Apparatus required for endotracheal intubation of a rabbit under direct vision, using either a laryngoscope (Wisconsin blade, size 1) (a) or an otoscope (b). (c) A 3 mm endotracheal tube with low dead space connector (d). (e) Paediatric introducer. (f) lignocaine spray. The apparatus illustrated would be suitable for rabbits weighing 3–4 kg.

carried out under direct vision (Figure 3.19). As with the pig and the cat, the larynx should be sprayed with lignocaine before intubation is attempted. When using an otoscope, after passing the introducer through the larynx the otoscope is removed, and only then is the endotracheal tube passed over the introducer and into the trachea. The introducer is then withdrawn and the tube tied in position.

An alternative technique for intubation does not require visualization of the larynx. The rabbit is placed in sternal recumbrancy, its head gripped firmly and extended, and lifted so that its forelegs are only just touching the operating table (Figure 3.20) The endotracheal tube is then advanced gently over the tongue towards the larynx. The operator listens for breath sounds at the end of the tube, or alternatively, if a clear polyethylene tube is used, looks for the presence of condensation. Loud breath sounds or condensation indicate that the tube tip is close to the larynx. As the rabbit breathes in, the tube is gently advanced. If it fails to enter the larynx, as indicated by cessation of breath sounds and loss of condensation, then the tube is withdrawn and another attempt made. This technique does not allow spraying of the vocal cords to prevent possible laryngospasm, but is easy to master, particularly in slightly larger rabbits (> 3 kg).

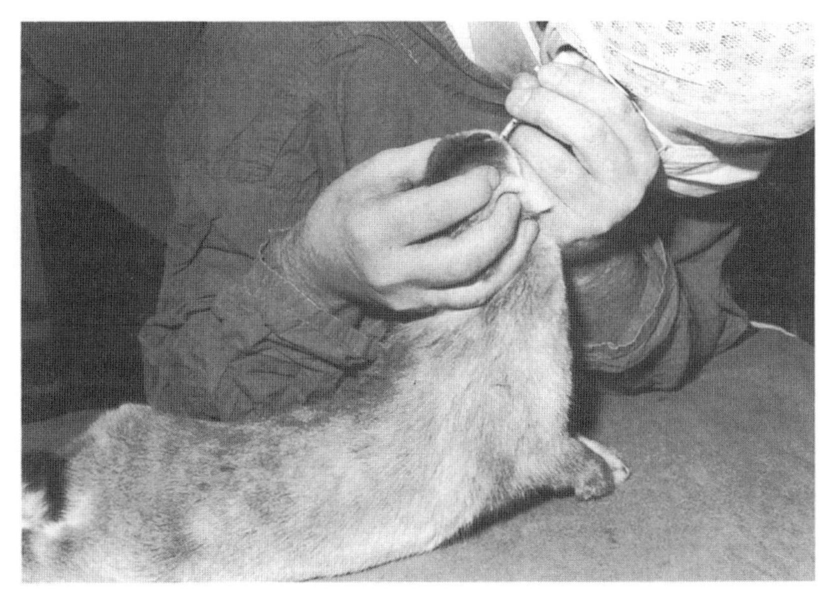

Fig. 3.20. Rabbit positioned for blind intubation. The position of the tip of the endotracheal tube is determined by listening to the respiratory sounds or by observing the condensation which appears in the tube lumen on each expiration.

iv. Rat. Intubation of the rat is possible using a purpose-made laryngoscope (Costa *et al.*, 1986) (Figure 3.21) or an otoscope. The rat is positioned on its back, and the tongue pulled gently forward and to one side. The laryngoscope or otoscope is then inserted until the larynx can be visualized. The animal may then be intubated using a suitably sized (16–12 gauge) arterial cannula (e.g. Abbocath, Abbott Laboratories Ltd, Appendix 7). Some modification of the Luer fitting is needed to provide connections to an appropriate anaesthetic circuit, and care must be taken to ensure that these connectors introduce only a minimum of dead space into the circuit. To avoid inadvertent intubation of one bronchus, and to provide a seal around the larynx, a small piece of rubber tubing can be positioned around the catheter, about 0.75–1 cm from the tip. Alternatively, some Micropore tape (3M) can be applied to make a similar cuff. This will reduce the leakage of gas around the tube, making ventilation more effective and will also improve the efficacy of positive end-expiratory pressure (PEEP) if this is required. A final modification which can be helpful is to Superglue a silk ligature onto the base of the Luer mount of the catheter, to enable it to be anchored to the rat's jaw.

When using an otoscope, it is necessary to use an introducer, since the

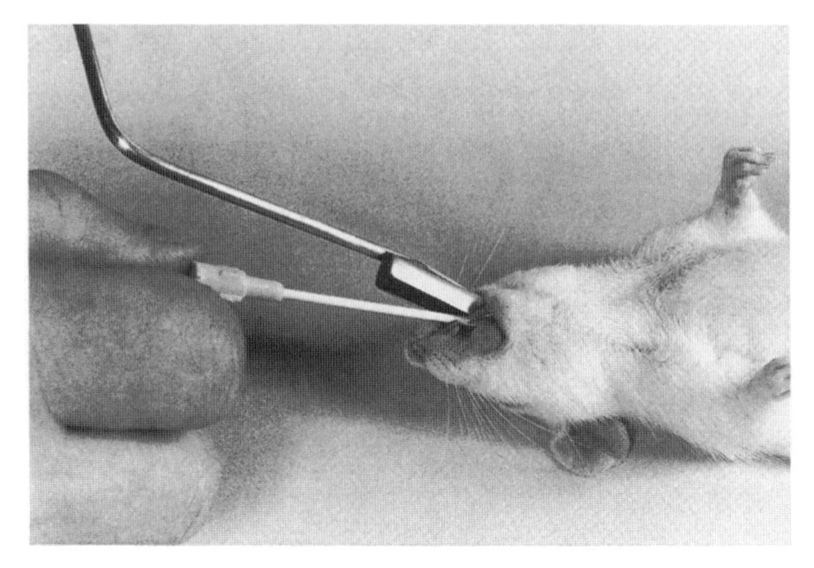

Fig. 3.21. Intubation of the rat using a purpose-made laryngoscope blade.

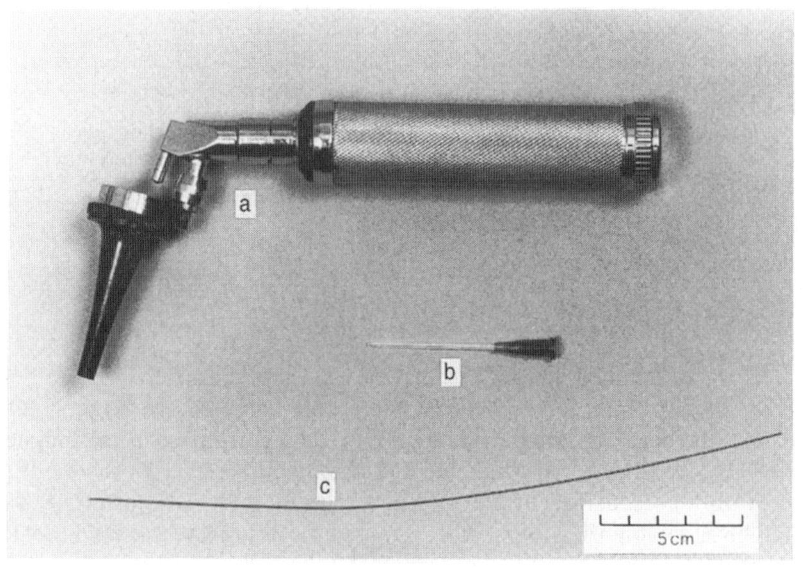

Fig. 3.22. Apparatus for intubation of the rat. (a) Otoscope. (b) 'Over the needle' catheter used as an endotracheal tube. (c) Seldinger guide wire. Note that the wire has been shortened and the 'J' end removed to ease handling and threading of the catheter.

cannula will not pass through the lumen of the otoscope. A guide wire from a Seldinger catheter makes an ideal introducer since its tip is soft and flexible (Figure 3.22). The wire is passed through the larynx under direct vision, the otoscope carefully removed, and the endotracheal tube threaded over the wire into the trachea. These wires can be purchased separately, and a 0.7 mm diameter wire will fit through both an 18 gauge and a 16 gauge catheter. Alternatively, the neck may be transilluminated using a powerful light source and the mouth opened using a small gag (Remie *et al.*, 1990). The tongue is pulled forward and a bright spot of light can be seen, which flashes as the rat breathes; this indicates the opening of the larynx. A combination of the last two techniques has proved most successful when training new research workers. An otoscope speculum is used to visualize the larnyx, which is illuminated via a powerful light source positioned over the neck. A Seldinger wire is inserted through the larynx into the trachea, the speculum is removed, a catheter threaded over the wire into the airway, and the wire removed.

v. Guinea pig, mouse, gerbil and hamster. Intubation of the mouse, gerbil or hamster is difficult and requires especial skill and purpose-made apparatus. A suitable set of laryngoscope blades has been described by Costa *et al.* (1986). The guinea pig can also be intubated using a purpose-designed laryngoscope blade, or the technique employing an otoscope (described above for the rat) can be used. As with the rat, use of an otoscope in combination with transillumination of the neck provides optimal conditions for intubation. Positioning of the otoscope is more difficult than in the rat, and a narrow speculum is needed to pass between the cheek teeth. The pharynx narrows markedly at the junction with the larynx and oesophagus, and considerable care must be taken to avoid inserting the speculum too far and occluding the larynx. As with the rat, intubation is achieved by passage of a Seldinger guide wire through the larynx, removing the otoscope, then passing a 16–12 gauge catheter over the wire into the trachea.

vi. Birds. Intubation of birds is relatively simple, since the opening to the airway is positioned much further anteriorly compared with mammals. Opening the beak enables the opening to be seen at the base of the tongue. Intubation is assisted by pulling the tongue forwards. Larger birds (poultry) can be intubated using standard paediatric tubes (2.5 mm upwards), but small birds require the use of either intravenous or urinary catheters, cut to a suitable length. Uncuffed endotracheal tubes should always be used; cuffed tubes can cause pressure necrosis of the tracheal mucosa because of the presence of complete tracheal cartilage rings in these species.

3. Drugs available

a. Inhalation anaesthetics. Information concerning the range of concentrations of the different inhalation anaesthetics which are required for induction and maintenance of anaesthesia is listed in Table 3.3. Several factors influence the apparent potency and efficacy of the different inhalation anaesthetics. The potency of each drug is indicated by its minimum alveolar concentration (MAC_{50}) value. This value, most commonly referred to simply as MAC, is the alveolar concentration of an anaesthetic required to block the response to a specified painful stimulus, e.g. clamping a haemostat onto a digit, in 50% of a group of animals. The lower the MAC value, the lower the concentration required, so the more potent the anaesthetic (Table 3.4).

The concentration of anaesthetic that can be delivered to the animal is influenced by the drug's boiling point. The lower the boiling point of an anaesthetic, the easier it is to vaporize, and so the higher the concentration that can be delivered. This is of considerable practical importance when selecting an anaesthetic agent and deciding how to vaporize it. A drug that is very potent, i.e. one with a low MAC value and also with a low boiling point, which makes it easy to vaporize, must be used with great care. There will be a considerable risk of overdosing the patient unless vaporization is carried out in a controlled way, using a calibrated vaporizer. Less potent anaesthetics, i.e. ones with higher MAC values and higher boiling points, can be used with greater confidence in simple apparatus, since dangerously high concentrations will not usually be produced.

The speed of induction of anaesthesia and the rate of recovery is affected

TABLE 3.3
Induction and maintenance concentrations of inhalation anaesthetic agents

Anaesthetic	Concentration for induction of anaesthesia (%)	Concentration for maintenance (%)	Minimum alveolar concentration (indicates relative potency of different agents) (in rat)
Enflurane	3–5	1–3	–
Ether	10–20	4–5	3·2
Halothane	4	1–2	0·95
Isoflurane	4	1·5–3	1·38
Methoxyflurane	3	0·4–1	0·22
Nitrous oxide	–	–	250

TABLE 3.4
Physical characteristics and relative potency (MAC_{50}) of different volatile anaesthetic agents

	Enflurane (Ethrane)	Halothane (Fluothane)	Isoflurane (Aerrane)	Methoxyflurane (Metofane)	Nitrous oxide
Vapour pressure (mmHg at 20°C)	172	242	240	23	Gas at room temperature
Vapour concentration (% saturated at 20°C)	23	32	32	3	100
MAC (in dog)	2.2	0.87	1.28	0.23	188–222
Stability in soda lime	Stable	Slight decomposition	Stable	Slight decomposition	Stable
Solubility:					
Blood/gas partition coefficient	2	2.5	1.4	15	0.5
Rubber/gas partition coefficient	74	120	62	630	1.2
Percentage of anaesthetic recovered as metabolite (in humans)	2.4	20–25	0.17	50	0.004

Data adapted from Steffey (1994).

TABLE 3.5
Minimum alveolar concentration (MAC_{50}) values (%) for inhalation anaesthetic agents in different species

	Ether	Halothane	Enflurane	Isoflurane	Nitrous oxide
Human	1.92	0.75	1.68	1.15	105
Primate	–	1.15	1.84	1.28	200
Dog	3.04	0.87	2.20	1.41	222
Pig	–	1.25	–	1.45	277
Sheep	–	–	–	1.58	–
Cat	2.10	0.82	1.20	1.63	255
Rat	3.20	1.10	2.21	1.38	150
Mouse	3.20	0.95	1.95	1.41	275
Rabbit	–	1.39	2.86	2.05	–

Data from Drummond (1985), Mazze *et al.* (1985) and Steffey (1994).

by the concentration of anaesthetic delivered, its anaesthetic potency (MAC value), and by its blood/gas partition coefficient. The partition coefficient influences the rate at which the concentration of anaesthetic in the brain approaches that necessary for anaesthesia to be produced. The higher the partition coefficient, the slower the rate of induction of anaesthesia and the slower the recovery rate. These properties are summarized in Table 3.4. The MAC values of anaesthetic agents are relatively constant between species (Table 3.5), with the exception of nitrous oxide (see Chapter 5).

Of particular concern to some research workers is the fate of inhalation anaesthetics once absorbed into the animal. A common misconception is that all of the agent which is inhaled is exhaled from the body. Many inhalation anaesthetics undergo significant metabolism, and this can result in induction of liver enzyme systems, as may occur following use of injectable anaesthetics. This can be of significance if the animal is to be used subsequently in a study which involves assessing the effects *in vivo* of a novel pharmaceutical or other compound. Although information is available concerning long-term exposure to inhalation anaesthetics (Linde and Berman, 1971; Brown and Sagalyn, 1974), there is little information concerning the effects on liver enzyme systems of brief periods of exposure. One means of avoiding this effect is to use isoflurane, an anaesthetic which undergoes virtually no metabolism (Eger, 1981). If other agents are used, it seems reasonable to suggest that brief periods

(< 5 min) of anaesthesia are unlikely to cause significant effects, but more prolonged exposure to anaesthetizing concentrations may result in induction of enzyme systems.

i. Operator safety. A variety of hazards have been reported as being associated with pollution of the operating room environment with anaesthetic gases (Ferstandig, 1978; Green, 1981a). The risk of explosion or fire associated with agents such as ether and cyclopropane is well recognized, and appropriate precautions must be taken to avoid these dangers. The risks to personnel that may arise from chronic exposure to low levels of certain inhalational anaesthetics are much more difficult to assess. The results of the many studies designed to determine these associated risks vary considerably but, at present, it would seem sensible to take appropriate steps to minimize operating theatre pollution.

A variety of gas-scavenging systems are available and it should be possible to obtain equipment suitable for most applications. A scavenging system suitable for small laboratory animals has been described by Hunter *et al.* (1984), and this, and others, are available commercially (International Market Supply; Viking Medical; Appendix 7). Note that systems using activated charcoal are not effective in removing nitrous oxide. Even with an effective scavenging system, spillage of waste gas will occur when the lid of an anaesthetic chamber is removed to gain access to the anaesthetized animal. If this is considered a significant problem, then either the whole procedure can be carried out in a fume hood, or specially designed chambers can be used, which completely remove the anaesthetic gases before the chamber is opened (see Figure 3.6).

b. Specific agents.

i. Ether (diethyl ether)

Desirable effects. Ether is easy to vaporize in simple apparatus. It is difficult to kill an animal with an overdose of ether, so it is a relatively safe agent for inexperienced anaesthetists.

Undesirable effects. Induction is unpleasant for the animal, and the irritant properties of ether can cause coughing, profuse bronchial and salivary secretions and occasional laryngospasm. Ether can cause pre-existing chronic respiratory disease to develop into an acute severe infection following recovery from anaesthesia and this may be particularly important in rodents and rabbits. Ether is flammable and forms explosive mixtures with both oxygen and air.

Special comments. Induction and recovery from anaesthesia are relatively slow. This is advantageous for inexperienced anaesthetists, as it makes inadvertent overdosage less likely. Conversely, a prolonged induction period can present problems in restraining the animal, particularly since most animals strongly resent inhaling the vapour.

Administration of ether stimulates catecholamine release (Carruba *et al.*, 1987) which counteracts the depressant effect that this anaesthetic exerts on the heart, so that blood pressure is maintained at near-normal levels at all except deep levels of anaesthesia. The catecholamine release also results in a moderate rise in blood glucose concentrations, and in a wide range of other metabolic changes, which may interfere with particular research protocols. Ether is not, as commonly believed, an inert compound. It undergoes metabolism and exposure to ether results in induction of liver enzyme activity (Linde and Berman, 1971)

Although ether is a popular anaesthetic, its use for induction is unpleasant for the animal and hazardous in several species, particularly guinea pigs. Its explosive properties make it a significant safety hazard: it is certainly inadvisable to kill animals with ether as the carcasses may be stored in refrigerators which are not spark-proof and an explosion may result. In its traditional role in anaesthetic chambers ('ether jars'), ether should be replaced with methoxyflurane.

ii. Halothane

Desirable effects. Halothane is easy to vaporize, and induction and recovery are rapid (1–3 minutes). It is a potent anaesthetic, is non-irritant, and is neither flammable nor explosive.

Undesirable effects. Halothane has a depressant effect on the cardiovascular system. Moderate hypotension is produced at surgical levels of anaesthesia because of a reduction in cardiac output and peripheral vasodilation. A dose-dependent depression of respiration also occurs. Some hepatic metabolism of halothane occurs and marked liver microsomal enzyme induction may follow anaesthesia (Wood and Wood, 1984).

Special comments. The desirable effects listed above make halothane an excellent agent for maintaining anaesthesia in most species. It should always be administered using a calibrated vaporizer since dangerously high concentrations can be attained if simple apparatus is used. Although recovery is usually rapid, it may be considerably delayed following prolonged, deep anaesthesia.

iii. Enflurane

Desirable effects. Induction and recovery from anaesthesia are rapid, so the depth of anaesthesia can be altered easily and rapidly. Enflurane is non-flammable, non-explosive and non-irritant.

Undesirable effects. Enflurane produces cardiovascular and respiratory depression, comparable to that occurring during halothane anaesthesia.

Special comments. Enflurane is largely eliminated via the lungs and, unlike halothane, very little drug is metabolized in the liver. This may offer advantages in certain experimental situations, although there is otherwise little to choose between halothane and enflurane in terms of efficacy as anaesthetic agents.

iv. Isoflurane

Desirable effects. Isoflurane produces very rapid induction and recovery from anaesthesia and the depth of anaesthesia can be altered easily and rapidly. It is non-irritant, non-explosive and non-flammable.

Undesirable effects. Isoflurane produces slightly more severe respiratory depression than does halothane, but slightly less depression of the cardiovascular system. Its pungent odour has been reported to cause breath-holding during induction in children, but this does not appear a significant problem in most species, with the exception of the rabbit (see Chapter 7, Rabbit section).

Special comments. The main advantage of using isoflurane in experimental animals is that it undergoes even less biotransformation than enflurane and is almost completely eliminated in exhaled air. This suggests that there will be little effect on liver microsomal enzymes and, hence, minimal interference in drug metabolism or toxicology studies (Eger, 1981). This characteristic, together with the rapid induction and recovery from anaesthesia, has led to the widespread adoption of isoflurane in many research establishments.

v. Methoxyflurane

Desirable effects. Methoxyflurane is non-irritant, non-flammable and non-explosive in air or oxygen. It has a potent analgesic effect and has some post-operative analgesic action.

Undesirable effects. Methoxyflurane produces some respiratory and cardiovascular system depression, but generally less than halothane at comparable depths of anaesthesia. Metabolism of methoxyflurane results in fluoride ion release which may cause renal damage. The significance of this hazard in animals is small except following very prolonged periods of anaesthesia (Murray and Fleming, 1972).

Special Comments. Because of the slow rate of induction with methoxyflurane, in larger species it is best used for maintenance of anaesthesia following induction with short-acting injectable agents. In small animals it can safely be used in anaesthetic chambers, where its slow induction and the low vapour concentration produced can be an advantage in reducing the risk of inadvertent overdose. It is an excellent agent for inducing and maintaining anaesthesia in neonatal animals.

vi. Sevoflurane and desflurane. Sevoflurane and desflurane are potent anaesthetic agents with characteristics similar to isoflurane. Desflurane (I–653) has been released for clinical use in humans, but requires a newly designed pressurized vaporizer to provide stable concentrations of the anaesthetic. It undergoes the least degree of metabolism (Koblin, 1992) and induction and recovery from anaesthesia are the most rapid of the volatile anaesthetics (Eger, 1992). Sevoflurane undergoes a similar degree of metabolism as isoflurane, and is unstable in the presence of soda lime, the carbon dioxide absorber used most commonly in closed circuit anaesthesia. The inital high cost of both agents, and the lack of many significant advantages compared with isoflurane, suggests that they are unlikely to be widely used in laboratory animal anaesthesia.

vii. Nitrous oxide.

Desirable effects. Nitrous oxide causes minimal cardiovascular and respiratory system depression.

Undesirable effects Nitrous oxide has very low anaesthetic potency and cannot be used alone to produce anaesthesia, or even unconsciousness, in most species. It reacts with vitamin B_{12}, producing vitamin depletion after prolonged anaesthesia (over 6 hours), and can cause bone marrow depression and blood dyscrasias.

Special comments. Since nitrous oxide has minimal effects on the respiratory and cardiovascular systems, it can be used to reduce the required concentration of other agents and so reduce the overall degree of depression of blood pressure or respiration at a particular depth of anaesthesia. It is usually administered as a 50:50 or a 60:40 mixture with oxygen. Following cessation of prolonged nitrous oxide administration, 100% oxygen should be administered to prevent so-called diffusion hypoxia. This phenomenon causes lowered alveolar oxygen tension due to the rapid diffusion of nitrous oxide from the blood to the alveoli. Because of its low analgesic potency, nitrous oxide must never be used as the sole anaesthetic agent in association with neuromuscular blocking agents such as pancuronium. Its main value lies in reducing the required concentration of other more potent agents which have more marked side-effects. It is important to note that nitrous oxide is not absorbed by the activated charcoal canisters in some gas-scavenging systems. If nitrous oxide is used, then an active scavenging system that ducts expired gases directly to the room ventilation extract must be used.

A common misconception is that it is necessary to adminster nitrous oxide in order to administer other inhalation anaesthetics. This is not the case, and all of the other agents mentioned above can safely be administered in 100% oxygen. It is only necessary to avoid this if prolonged periods of anaesthesia are planned (> 24 hours) when the inspired oxygen concentration should be reduced (to approximately 40%) to avoid the possible development of oxygen toxicity. This can be achieved without the use of nitrous oxide by using an air/oxygen mixture or nitrogen/oxygen mixture, the other gas being supplied from a compressed gas cylinder. If the gases are mixed at the outlet from the anaesthetic machine, then the delivered concentration of anaesthetic vapour will be reduced, and the vaporizer setting should be increased accordingly.

viii. Older agents. A number of other anaesthetic agents have been used, but they are primarily of historical interest only. Chloroform has numerous side-effects which resulted in it being discarded from human and veterinary clinical practice, and which make it unsuitable for laboratory use. Trichloroethylene produces good analgesia, is inexpensive, non-inflammable and non-explosive. It also causes only minimal cardiovascular system depression but has poor muscle relaxant properties, low anaesthetic potency, and decomposes in the presence of soda lime to form toxic and explosive products, so that it must never be used in closed circuits. Trichloroethylene undergoes extensive hepatic metabolism and is rarely used for animal anaesthesia. Cyclopropane can be used to produce rapid induction of anaesthesia and the depth of anaesthesia can be altered smoothly and rapidly. Recovery from anaesthesia is also rapid. Unfortunately, cyclopropane is flammable and explosive in air and oxygen, and this hazard limits its use in most laboratories.

B. Administration of anaesthetic agents by injection

1. Equipment

Although the equipment required for injection of anaesthetic drugs consists basically of a syringe and needle, some attention should be given to the range of syringe and needle sizes available and also to the use of indwelling catheters, cannulae, extension tubing and infusion devices.

a. Syringes. Plastic disposable syringes are to be preferred to glass re-usable syringes. Single-use syringes should not be re-sterilized for further use. Ensure that a syringe of appropriate volume is used so that the required dose of anaesthetic can be administered accurately. The syringes designed for insulin administration to human patients are particularly useful for administering small doses of drugs to rodents (Figure 3.23). Select a syringe design that is comfortable to hold and that enables a firm grip to be maintained even when the barrel is wet. Avoid using syringes that have been stored for longer than the time recommended by the manufacturers, as the plastic may have become brittle and can fracture during use.

b. Needles and cannulae. Disposable hypodermic needles should be used and an appropriate gauge selected for each purpose. Needles should never be re-sterilized as they rapidly become blunt when used, and injection with a blunt needle can cause considerable discomfort. Successful venepuncture of small vessels is particularly difficult to achieve if the needle has been blunted; for this reason it is advisable to replace the needle after drawing up liquid from a rubber-capped vial.

Often, it is preferable to use a butterfly-type infusion set, rather than a simple hypodermic needle. These infusion devices provide a short length of flexible catheter between the needle and the syringe so that movements of the animal during injection are less likely to result in the needle becoming dislodged from the vein (Figure 3.23). This is particularly important when inducing anaesthesia with short-acting barbiturates such as methohexitone, since administration of an inadequate dose of drug may produce involuntary excitement. If the ensuing limb movements result in displacement of the needle from the vein, its replacement may be virtually impossible.

If successive intravenous injections of anaesthetic are required, it is preferable to use an indwelling cannula rather than a butterfly-type infusion line or a hypodermic needle. A flexible cannula will not pierce the vessel wall should movements of the animal occur, so inadvertent extravascular injection will be avoided. Several types of cannula are available:

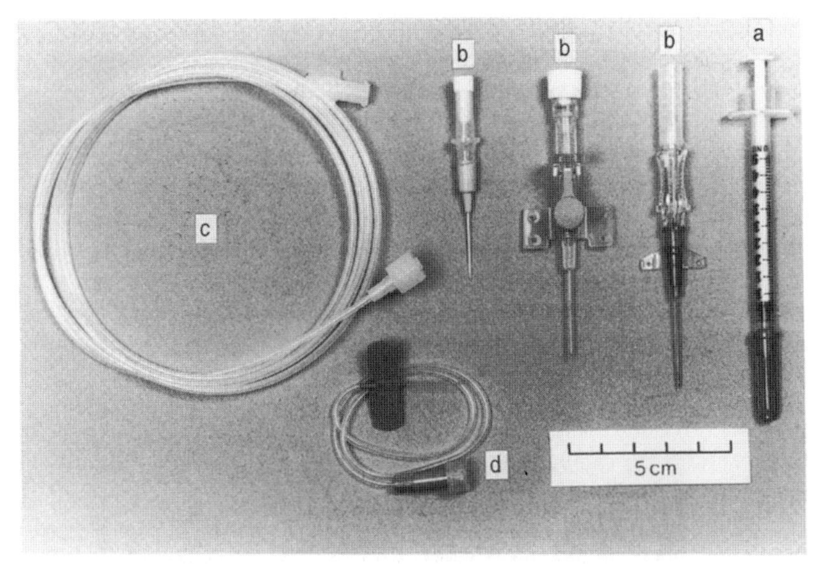

Fig. 3.23. Apparatus for intravenous injection and intravenous infusions in small animals. (a) Insulin syringe (0.5 ml volume, 25 G needle). (b) Over-the-needle catheters (24–21G). (c) Anaesthetic extension tubing. (d) 'Butterfly' style infusion set.

they can be broadly grouped as over-the-needle designs, in which the flexible cannula is placed on the outside of a needle which acts as an introducer, and through-the-needle designs, in which the cannula runs through the needle. In most circumstances over-the-needle cannulae are preferable for use in small animals since they allow the largest possible cannula to be inserted into the vessel (Figure 3.23). In large animals, the skin may offer significant resistance to passage of the cannula and may damage an over-the-needle type, but this does not occur when using a through-the-needle design. An alternative solution is to make a very small skin incision with a scalpel blade to allow easy passage of an over-the-needle cannula.

Although such cannulae are relatively expensive the advantages of maintaining a secure route for intravenous drug administration can be considerable. In addition to administration of anaesthetics, other drugs and intravenous fluids can be administered rapidly even by relatively unskilled assistants. It is important that the cannula is securely anchored in place. This can be achieved as illustrated in Figure 3.24. When anchoring catheters in the marginal ear vein in species such as the rabbit or sheep, it is helpful to cut off one wing to reduce the risk of dislodging the catheter.

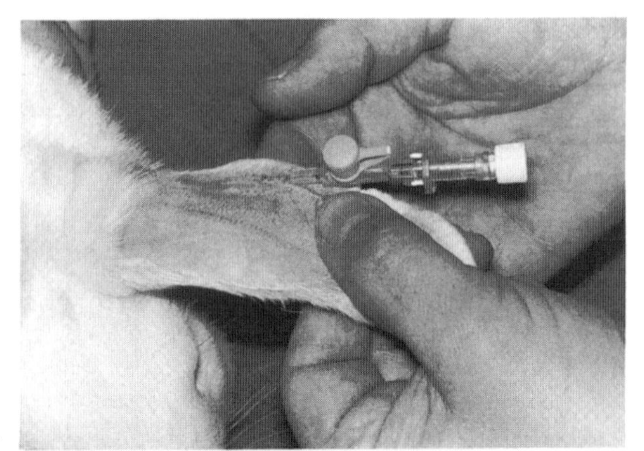

a

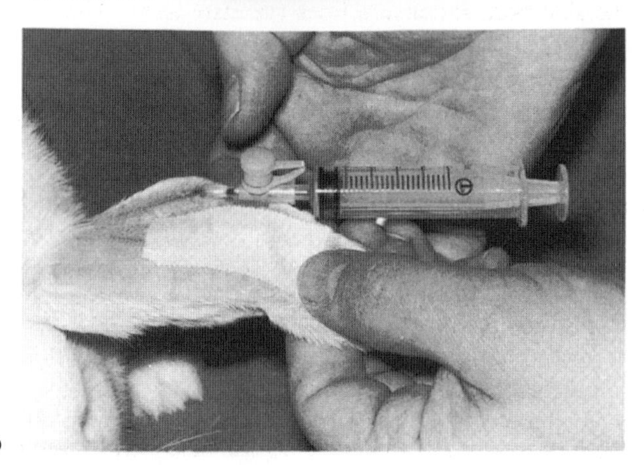

b

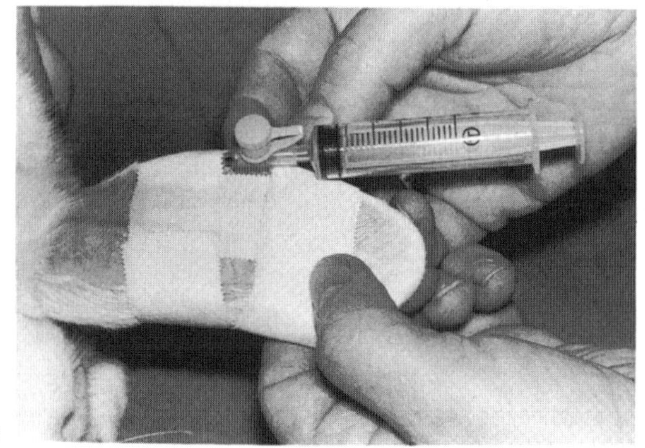

c

c. Extension lines. It is often inconvenient to require access to the site of cannulation for repeated drug administration and this can be avoided by the use of a suitable length of plastic tubing as an extension cannula (see Figure 3.23). Lines equipped with a Luer-locking fitting are preferable as they are less likely to become disconnected. A problem with many of the extension lines produced for human use is their large volume, which can cause problems if different drugs are to be administered successively to the animal. It is often undesirable to administer a bolus of 4–5 ml of saline to a small animal to flush an infusion line and this drawback should be considered when selecting apparatus. Low-volume extension lines (< 1.0 ml) are available from Vygon (UK) Ltd (Appendix 7) (Figure 3.23). A useful compromise is to select an indwelling cannula with a side injection port (Figure 3.23). Routine infusion of anaesthetic can be carried out through an extension line and administration of other drugs through the side port. Extension lines are also useful when administering large volumes of drug by the intramuscular route to larger animals such as pigs. Use of an extension between the needle and syringe enables placement of the needle, followed by controlled injection without the need to restrain the animal.

d. Infusion pumps. It is often convenient to administer intravenous anaesthetics by continuous infusion. A range of infusion pumps is available commercially and the cost of sophisticated, microprocessor-controlled models has fallen rapidly. Pumps designed for clinical use in humans generally operate using a 50 ml syringe. Although this syringe size is somewhat excessive for use in small animals, the rate at which drugs can be delivered can be as little as 0.1 ml h^{-1}.

Smaller volume pumps, particularly those designed for insulin infusion, are suitable for use in small animals. Purpose-designed infusion pumps that allow the use of different syringe sizes are more versatile, and are a worthwhile investment if total intravenous anaesthesia is to be employed.

If an infusion pump is not available, drugs can be administered using an intravenous infusion set and a burette to allow better control over the volumes administered. The use of such gravity-feed devices has the obvious disadvantage that changes in the position of the cannula, or movements of the limb, can greatly affect the infusion rate. Nevertheless, such simple devices can be used successfully, particularly if a

Fig. 3.24. Placement and anchoring of an intravenous catheter in the rabbit. (a) The hair overlying the marginal ear vein has been clipped, and EMLA cream applied to prevent any response to venepuncture. One wing has been removed from the catheter which is introduced into the vein. The needle is withdrawn slightly into the catheter lumen before advancing further into the vessel. (b) A strip of tape is placed along the length of the ear, anchoring the wing of the catheter. (c) Two additional strips of tapes are placed around the ear, above and below the catheter wing.

central venous cannula (which is less susceptible to occlusion) is used. The apparatus available for infusion of anaesthetics has been reviewed by Glen (1988). Further details of infusion techniques are given in Chapter 5.

2. *Routes of administration*

Injectable anaesthetic agents can be administered by a variety of routes. Intravenous administration is usually preferable, since this produces the most predictable and rapid onset of action. This enables the drug to be administered to effect to provide the desired depth of anaesthesia. Practical considerations, such as the absence of suitable superficial veins or difficulty in providing adequate restraint of the animal, may limit the use of this route in some laboratory species. Administration by intramuscular (i/m), intraperitoneal (i/p) or subcutaneous (s/c) injection is relatively straightforward in most species, but the rate of drug absorption, and hence its anaesthetic effects, may vary considerably. It is important to appreciate the very great variation in response to anaesthetics that occurs between animals of different strains, age, and sex. The magnitude of these effects is illustrated

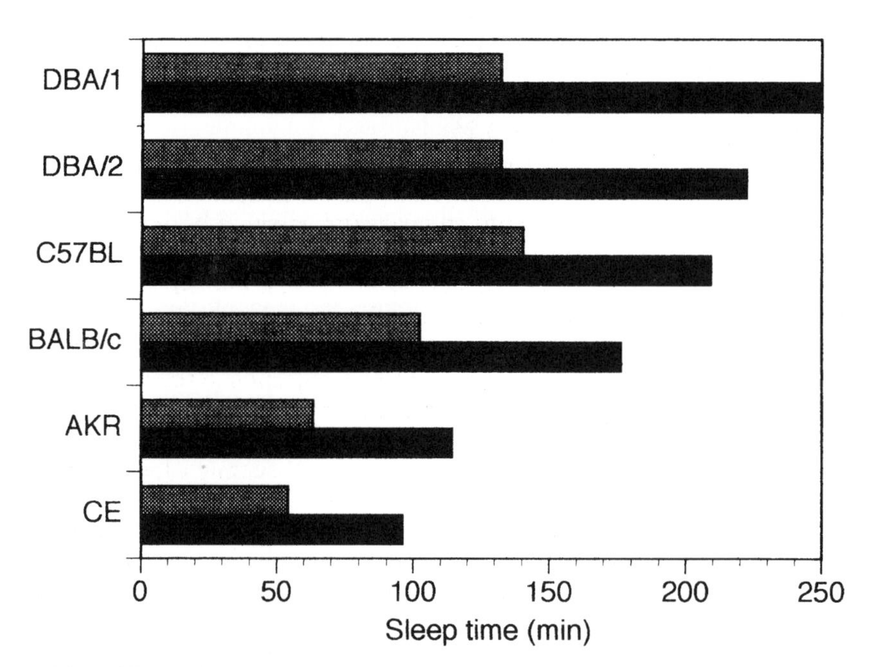

Fig. 3.25. Variation in sleep times (duration of loss of righting reflex) following a standard dose of pentobarbitone (60 mg kg^{-1}) in a range of different inbred strains of mouse. Note also the marked sex difference in response: male mice, black bars, female, shaded bars. Data from Lovell, 1986b.

with pentobarbitone in mice in Figure 3.25, but this variation must be anticipated in all species and with all anaesthetic drugs. When using an anaesthetic regimen for the first time, it is essential to assess its effects in one animal, before beginning to anaesthetize the remainder of the group. This will enable the recommended doses to be adjusted to suit the responses of the particular animals being used. As mentioned above, this variability in response can be a particular problem in small rodents, since most injectable anaesthetics are administered to these species by the i/p route as a single dose. When administering anaesthetic drugs in this way it is impossible to adjust the dose according to the individual animal's response, so inadvertent overdosing and underdosing will frequently occur, until experience is gained with a particular strain, age and sex of animal. Variation in response also occurs with changes in environmental factors, and standardization of all of these variables will not only simplify anaesthetic dose calculations, but also constitutes good experimental design. When selecting anaesthetic agents for intramuscular, intraperitoneal or subcutaneous administration, it is also advisable to select those which have a wide safety margin.

A further disadvantage of the intraperitoneal or intramuscular route is that relatively large doses of anaesthetic must be given to produce the required effect. Absorption is slow relative to intravenous administration, residual drug effects can persist for prolonged periods, and so full recovery can be very prolonged (Figure 3.26).

An additional consideration with i/m, i/p or s/c injection is that injection of an irritant compound can cause unnecessary pain or discomfort to the animal. This is a particular concern with intramuscular injection in small rodents, and there are a number of reports of tissue reactions and myositis (Gaertner *et al.*, 1987; Smiler *et al.*, 1990; Beyers *et al.*, 1991). For this reason it is recommended that the intramuscular route is avoided in small rodents.

Intravenous administration avoids the problems discussed above, and the technical problems associated with intravenous injection in small mammals are often more imagined than real. Research workers may avoid intravenous anaesthesia, and yet administer other compounds by the intravenous route as part of their research protocol. Before discounting administration, consider whether the necessary expertise is already available, or if developing this expertise would be worthwhile. In rats, for example, placement of an over-the-needle catheter allows both intravenous anaesthesia and administration of other drugs and fluids as necessary. A number of short-acting anaesthetics can be used to provide 5–10-minute periods of anaesthesia (see below and Chapter 7), and some are suitable for continuous infusion to provide long-term anaesthesia. When carrying out venepuncture in animals, consider using lignocaine/prilocaine cream (EMLA, Astra) to produce anaesthesia of the skin (see Chapter 2).

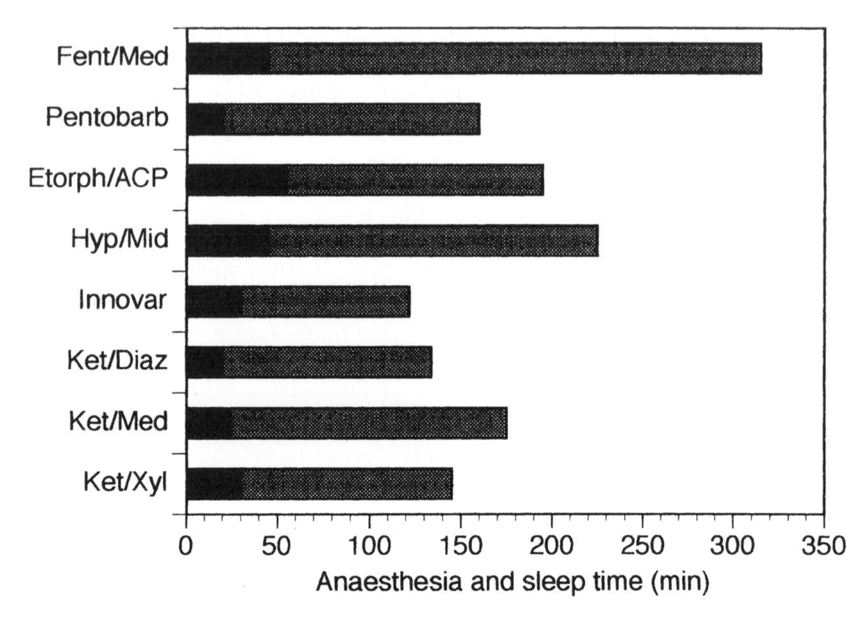

Fig. 3.26. Typical anaesthesia times (black) and sleep times (shaded) in laboratory rats anaesthetized with injectable anaesthetic agents: fentanyl/medetomidine (Fent/Med), pentobarbitone (pentobarb), etorphine/acepromazine (Etorph/Ace), fentanyl/fluanisone/midazolam (Hyp/Mid), fentanyl/droperidol (Innov), ketamine/diazepam (Ket/Diaz), ketamine/medetomidine (Ket/Med), ketamine/xylazine (Ket/Xyl). Note that very considerable variation can occur between different strains (see Fig. 3.25).

3. Drugs available

a. Barbiturates

i. Pentobarbitone

Desirable effects. Pentobarbitone can be administered by either intravenous or intraperitoneal injection and can be used in a wide range of animal species.

Undesirable effects. Severe cardiovascular and respiratory system depression is produced and this drug has poor analgesic activity. Recovery can be prolonged, particularly after administration of an additional dose to prolong anaesthesia.

Special comments. Pentobarbitone is probably the most widely used laboratory animal anaesthetic. Surgical anaesthesia is attained in most small laboratory animals only when dosages close to those that cause respiratory failure have been administered. At these dose rates severe cardiovascular and respiratory depression are produced. Slow intravenous administration of a dose sufficient to produce basal narcosis, followed by further incremental doses, usually achieves surgical levels of anaesthesia reasonably safely.

Intraperitoneal administration of the calculated amount of drug as a single bolus is often associated with a high mortality rate because the anaesthetic dose is very close to the lethal dose, and there is considerable between-strain variation. Pentobarbitone is probably best used to provide hypnosis rather than anaesthesia and in most circumstances safer and more effective agents are available.

ii. Thiopentone

Desirable effects. Thiopentone produces smooth and rapid induction of anaesthesia following i/v injection and can be used in virtually all species.

Undesirable effects. Thiopentone has poor analgesic activity and causes transient apnoea after i/v injection. It is irritant if injected perivascularly. Repeated administration results in very prolonged recovery time (see Chapter 5).

Special comments. Thiopentone is a short-acting barbiturate which is useful for rapid induction of anaesthesia when administered intravenously. It is unstable in aqueous solution, so once reconstituted it should be used within 7–10 days. Its duration of action depends upon both the amount of drug injected and on the rate of injection. The doses quoted in Chapter 7 should be administered as follows: half the calculated dose should be given rapidly, followed by the remainder to effect over 1–2 minutes. This will result in 5–15 minutes of anaesthesia. Transient apnoea usually follows administration, but assisted ventilation is rarely required. Thiopentone solution is extremely irritant if injected perivascularly and should be diluted as much as practicable (preferably to enable use of a 1.25–2.5% solution). If extravascular administration occurs, then the area should be infiltrated with a solution of 1 ml of lignocaine 2% in 4 ml normal saline (Trim, 1987). Since thiopentone is highly irritant, primarily because of its high pH, it is not advisable to administer this agent by the intraperitoneal route if the animal is intended to recover from the anaesthetic. The drug's major use is by intravenous injection to provide rapid induction of anaesthesia, followed by maintenance using inhalational agents.

iii. Methohexitone

Desirable effects. Methohexitone produces smooth and rapid induction of anaesthesia after i/v administration and can be used in a wide range of species.

Undesirable effects. Like other barbiturates, methohexitone has poor analgesic activity and transient apnoea often occurs after induction. Recovery is frequently accompanied by muscular tremors unless suitable pre-anaesthetic medication has been administered.

Special comments. Methohexitone has a shorter duration of action than thiopentone and is about twice as potent. It should be administered as described above for thiopentone. Anaesthesia lasts for 2–5 minutes and several incremental doses can usually be given without unduly prolonging the rate of recovery. Methohexitone is a valuable drug for induction of anaesthesia provided that intravenous administration is possible. Although intraperitoneal administration has been reported, use of this route of administration often has less predicatable effects, with some animals failing to become anaesthetized.

iv. Inactin

Desirable effects. Inactin produces smooth induction after i/v administration and has a prolonged duration of action.

Undesirable effects. It produces variable analgesic activity.

Special comments. Inactin is a thiobarbiturate which has been claimed to produce prolonged anaesthesia in rats following intraperitoneal (Buelke-Sam *et al.*, 1978) or intravenous (Walker *et al.*, 1983) administration. Whilst it appears to be a satisfactory induction agent when given intravenously (resembling thiopentone in its effects), its effects when given by the intraperitoneal route may vary. In the author's experience, some rats remain lightly anaesthetized for several hours, whereas others appear completely recovered in under 60 minutes. Provided that an appropriate dose rate has been established in the particular strain of rat which is to be anaesthetized, Inactin can be used to produce prolonged anaesthesia.

b. Steroid anaesthetic agents

i. Alphaxalone/alphadolone

Desirable effects. Alphaxalone/alphadolone produces smooth induction of anaesthesia following i/v administration. Administration of repeated doses of the drug has little effect on recovery time. The solution is non-irritant and the compound has a wide margin of safety in most species (Child *et al.*, 1971, 1972b, c)

Undesirable effects. The solubilizing agent present in the commercial preparation of this anaesthetic promotes histamine release in dogs and the drug should not be used in this species. Mild histamine release also occurs frequently in cats, causing oedema of the paws, muzzle and ears.

Special comments. The commercial preparation of this anaesthetic agent consists of a mixture of two steroids, alphaxalone and alphadolone, together with a solubilizing agent, Chremophor EL (polyoxyethylated castor oil). Alphaxalone and alphadolone differ slightly in their anaesthetic potency, but dose rates of the commercial preparation are conventionally reported as mg kg^{-1} of total steroid.

Administration of alphaxalone/alphadolone by the intramuscular or intraperitoneal routes has very variable effects. Occasionally, light surgical anaesthesia is produced, but in most species the volume of drug required precludes intramuscular administration and absorption following intraperitoneal injection is very unpredictable (Green *et al.*, 1978). It is an effective agent for immobilizing small primates when administered intramuscularly, however. Following intravenous administration it produces rapid-onset anaesthesia followed by rapid recovery. The agent is non-irritant, and inadvertent extravascular injection does not appear to be associated with any adverse effects.

Since the drug is rapidly metabolized, it is an excellent agent for maintenance of long-term anaesthesia, although moderate hypotension may occur (Child *et al.*, 1972a; Dyson *et al.*, 1987). Continuous intravenous infusion can be used to provide safe and stable anaesthesia in sheep, pigs, primates, cats and rodents, although in larger species economic considerations may limit its usefulness. In rabbits, the degree of analgesia produced is insufficient for major surgery until large doses have been administered and at these dosages respiratory arrest often occurs.

Alphaxalone/alphadolone must not be used in conjunction with barbiturates. Although structurally related to the steroid hormones, alphaxalone and alphadolone have no significant endocrine effects (Child *et al.*, 1972b).

c. Dissociative anaesthetic agents

i. Ketamine

Desirable effects. Ketamine produces immobility in most species and can be administered by the intramuscular, intraperitoneal and intravenous routes. It causes only moderate respiratory depression in most species (note the comments on rodents, below) and produces an increase in blood pressure.

Undesirable effects. Skeletal muscle tone is increased. The degree of analgesia produced is very variable and in small rodents severe respiratory depression is produced following administration of the high dose rates needed for surgical anaesthesia. Recovery can be prolonged and may be associated with hallucinations and mood alterations (Wright, 1982).

Special comments. Ketamine produces a state of cataleptic sedation with apparent lack of awareness of the surroundings (White *et al.*, 1982). In species in which profound analgesia appears to be produced, spontaneous movements often occur but these are usually unrelated to surgical stimuli. In some species the corneal blink reflex is lost for prolonged periods and drying of the cornea may occur unless the eye is filled with a bland ophthalmic ointment as a preventive measure. Laryngeal and pharyngeal reflexes are maintained at all except very high dose rates, although salivary secretions are increased and airway obstruction remains a significant hazard. Ketamine is the drug of choice for immobilization of large primates, and it is an effective chemical restraining agent in cats and pigs and, to a lesser extent, in rabbits. Its effects in rodents are variable and high dose rates may be necessary to produce surgical anaesthesia (Green *et al.*, 1981a). It is extremely useful when administered in combination with medetomidine, xylazine or diazepam for the production of surgical anaesthesia in sheep, primates, cats, dogs, rabbits and small rodents (see Chapter 7). It is important to appreciate that the stimulatory effects of ketamine on the cardiovascular system do not offset the depressant effects of drugs such as xylazine, and use of these combinations almost invariably results in significant hypotension (Figure 3.27) (Middleton *et al.*, 1982; Allen *et al.*, 1986). Ketamine can be mixed with medetomidine, xylazine or acepromazine and the combination administered as a single injection. In all species it is preferable to use atropine or glycopyrrolate together with ketamine to reduce the otherwise excessive bronchial and salivary secretions which are produced. The chronic administration of ketamine results in

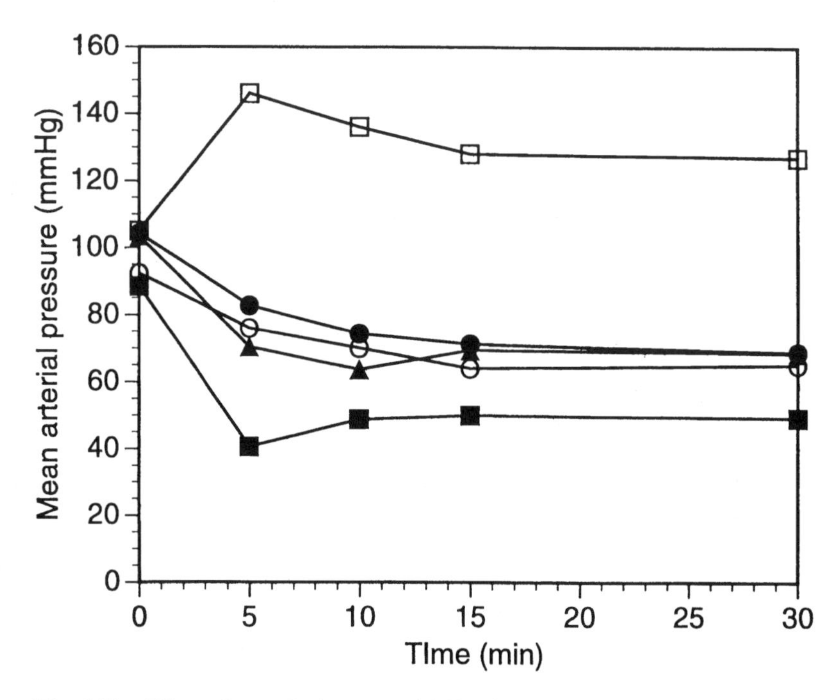

Fig. 3.27. Effect of anaesthesia on arterial blood pressure in the cat. Ketamine (open squares) alone causes an increase in blood pressure, however the addition of xylazine (solid circles) or acepromazine (triangles) results in hypotension, comparable to that produced by alphaxalone/alphadolone (open circles). Halothane (solid squares) causes marked hypotension. Data from Middleton *et al.* (1982), Allen *et al.* (1986), Dyson *et al.* (1987) and Ingwerson *et al.* (1988).

hepatic enzyme induction and this may decrease the efficacy of the agent on subsequent administrations (Marietta *et al.*, 1975).

d. Neuroleptanalgesic combinations

i. Fentanyl/fluanisone, fentanyl/droperidol, etorphine/methotrimeprazine, etorphine/acepromazine

Desirable effects. Neuroleptanalgesic combinations produce profound analgesia and can be administered by the intramuscular, intraperitoneal or intravenous routes to most species. The effects of these drug combinations can be reversed by administration of mu-opioid antagonists such as naloxone or nalbuphine (see below).

Undesirable effects. Neuroleptanalgesic combinations produce moderate or severe respiratory depression and a poor degree of muscle relaxation (see below). Hypotension and bradycardia may also be produced.

Special comments. Neuroleptanalgesic combinations consist of a potent narcotic analgesic, which can abolish the perception of pain, and a neuroleptic — a tranquillizer/sedative (e.g. acepromazine or fluanisone) — which suppresses some of the undesirable side-effects of the narcotic such as vomiting or excitement. The analgesics used in commercially available neuroleptanalgesic combinations are fentanyl and etorphine. When neuroleptanalgesic combinations are used alone, the adverse effects mentioned above can be marked and the poor degree of muscle relaxation produced may preclude their use for anything other than superficial surgery. When given in combination with a benzodiazepine (e.g. midazolam or diazepam), the dose of some neuroleptanalgesic mixtures can be reduced by 50–70% and the benzodiazepine produces good skeletal muscle relaxation. Used in this way, combinations such as fentanyl/fluanisone and midazolam are often the anaesthetic method of choice for rodents and rabbits. Although pharmacologically similar, fentanyl/fluanisone (Hypnorm) and fentanyl/droperidol (Thalamonol, Innovar Vet) differ in their effects in animals. As mentioned above, fentanyl/fluanisone in combination with diazepam or midazolam produces good surgical anaesthesia. The effects of a comparable mixture of fentanyl/droperidol and midazolam are much less predictable, and this latter combination cannot be recommended (author's unpublished observations; Marini *et al.*, 1993). One other commercially available combination, etorphine/methotrimeprazine (Immobilon SA) has been evaluated in combination with midazolam. Although surgical anaesthesia is produced, respiratory depression can be severe (Whelan and Flecknell, 1994, 1995).

An important advantage of these drug combinations is that their action is readily reversible by administration of narcotic antagonists such as naloxone, mixed agonist/antagonists such as nalbuphine, or partial agonists such as buprenorphine (Flecknell *et al.*, 1989a) (see Chapter 7). Fentanyl/fluanisone and fentanyl/droperidol are useful for providing restraint and analgesia for minor procedures, and the combination of fentanyl/fluanisone/midazolam is recommended for surgical anaesthesia in rodents and rabbits. The use of other neuroleptanalgesic combinations has been described (Green, 1975).

e. Other opioid combinations. Because of their potent analgesic action, short-acting opioids such as fentanyl and alfentanil can be used in combination with a variety of compounds to produce balanced anaesthesia. Mixtures of fentanyl or alfentanil and a benzodiazepine produce effective surgical anaesthesia in dogs (Flecknell *et al.*, 1989b) and pigs, and can be added to anaesthetic regimens in which analgesia would otherwise be inadequate (e.g. propofol) (Michalot *et al.*, 1980; Flecknell *et al.*, 1990b). Use of opioids often enables the production of profound analgesia without major effects on the cardiovascular system. Severe respiratory depression can occur when using high doses of opioids, although this can be overcome by the use of intermittent positive-pressure ventilation.

i. Fentanyl/medetomidine. Fentanyl and medetomidine can be combined to produce anaesthesia in dogs, rabbits, guinea pigs and rats. The combination is most effective in dogs and rats. In the dog the drugs are given by intravenous injection, in the rat the two compounds are combined and given as a single intraperitoneal injection.

Desirable effects. Surgical anaesthesia is reliably produced, with good muscle relaxation in some species (see Chapter 7). Anaesthesia is completely reversible by administering specific antagonists (nalbuphine or butorphanol and atipamezole).

Undesirable effects. Mild to moderate respiratory depression is produced. In the rat the relatively large volume for injection is inconvenient for the operator, but does not appear distressing to the animal. In the mouse, the combination causes urinary retention which may result in rupture of the bladder, and should not be used in this species.

Special considerations. The rapid and complete reversal of anaesthesia avoids the problems that may be associated with managing animals during the prolonged recovery that can be associated with other injectable anaesthetic regimens. Reversal of the fentanyl with a mixed opioid agonist/antagonist results in maintenance of post-operative analgesia.

ii. Etomidate and metomidate

Desirable effects. Etomidate and metomidate are short-acting hypnotics with minimal effects on the cardiovascular system.

Undesirable effects. There is little analgesic action when these drugs are used alone. Suppression of adrenocortical function occurs following prolonged infusion of etomidate (Kruse-Elliott *et al.*, 1987).

Special comments. Etomidate has been shown to cause litle cardiovascular depression in animals (Nagel *et al.*, 1979; Kissin *et al.*, 1983) and so may be useful as part of balanced anaesthetic regimens. Metomidate and etomidate are useful for providing unconsciousness (and therefore restraint) in many mammals, birds, reptiles and fish (Janssen *et al.*, 1975). Metomidate in combination with fentanyl, administered as a subcutaneous injection, is an effective anaesthetic combination for small rodents (Chapter 7) (Green *et al.*, 1981b).

f. Other anaesthetic agents

i. Propofol

Desirable effects. Rapid induction of a short period of anaesthesia occurs in a wide range of species. Recovery is smooth and rapid with little cumulative effect if additional doses are administered.

Undesirable effects. Analgesia is insufficient for major surgery in some species. A short period of apnoea occurs after induction and respiratory depression can occur with high doses of propofol. Prolonged infusion causes lipaemia because of the formulation (see below).

Special considerations. Propofol (di-isopropofol or ICI 35868) is an alkyl phenol (Glen, 1980; Glen and Hunter, 1984) which, because of its poor water-solubility, is prepared as an emulsion formulation in soya bean oil and glycerol. Intravenous administration of this compound produces rapid-onset anaesthesia in a wide range of species with a sleep time similar to thiopentone (Glen, 1980). In contrast to thiopentone, animals recover more rapidly following propofol administration, and sleep times are not greatly prolonged following repeated dosing (Glen, 1980).

Because of its rapid redistribution and metabolism, propofol must be given by rapid intravenous injection to be effective, otherwise the rapid redistribution to body tissues that occurs will prevent anaesthetic concentrations being achieved in the brain. Propofol produces a moderate fall in systolic blood pressure, and a small fall in cardiac output (Sebel and Lowdon, 1989). Propofol causes significant respiratory depression in most species, manifested either as a reduction in respiratory rate (Glen, 1980), or little change in rate but a fall in arterial oxygen tension, suggesting a fall in tidal volume (Watkins *et al.*, 1987). It is therefore advisable to provide supplemental oxygen. Propofol is believed to have no significant effects on hepatic function (Robinson and Patterson, 1985) or renal function (Stark *et al.*, 1985), or on platelet function or blood coagulation (Sear *et al.*, 1985). In humans, propofol causes a fall in intraocular pressure (Vanacker *et al.*, 1987), but no data are available for its effects in animal species. The pain on injection of propofol which has been reported in humans does not appear to be a significant problem in animals (Brearley *et al.*, 1988; Flecknell *et al.*, 1990b; Weaver and Raptopoulos, 1990). Propofol is non-irritant when injected perivascularly (Morgan and Legge, 1989).

ii. Tribromoethanol

Desirable effects. Tribromoethanol (Avertin) produces surgical anaesthesia in rats and mice, with good skeletal muscle relaxation and only a moderate degree of respiratory depression.

Undesirable effects. If incorrectly stored, or administered more than once, tribromoethanol is irritant to the peritoneum (see below).

Special comments. Tribromoethanol is administered by the intraperitoneal route, but decomposition of stored solutions can result in severe irritation and peritoneal adhesions following its use. Even if a freshly prepared solution is used, administration of a second anaesthetic at a later date is often associated with a high mortality rate (Norris and Turner, 1983). When properly prepared and stored, this anaesthetic is safe and effective in mice (Papaioannou and Fox, 1993).

ii. Chloral hydrate

Desirable effects. Chloral hydrate produces medium-duration (1–2 hours), stable, light anaesthesia (Sisson and Siegal, 1989; Field *et al.*, 1993). The drug has minimal effects on the cardiovascular system and on baroreceptor reflexes.

Undesirable effects. Chloral hydrate has poor analgesic properties and the large doses required for surgical anaesthesia can produce severe respiratory depression. Intraperitoneal administration to rats has been associated with a high incidence of post-anaesthetic ileus (dilation and stasis of the bowel) (Fleischman *et al.*, 1977). Although use of low concentrations of chloral hydrate (36 mg ml^{-1}) may reduce the incidence of this effect, it may not completely eliminate the problem.

Special comments. Chloral hydrate can often be replaced by more effective anaesthetics if surgical procedures are to be undertaken. As with many other anaesthetics, there is considerable strain variation in the response to chloral hydrate in rodents, and it is important to evaluate the drug's efficacy and safety in the particular strain of animals that will be used.

iv. *Alpha-chloralose*

Desirable effects. Alpha-chloralose produces stable, long-lasting (8–10 hours) but light anaesthesia. It produces minimal cardiovascular and respiratory system depression (Holzgrefe *et al.*, 1987; Svendsen *et al.*, 1990).

Undesirable effects. Alpha-chloralose has poor analgesic properties, although this varies considerably between different species and strains of animals. Both induction and recovery can be very prolonged and associated with involuntary excitement.

Special comments. Alpha-chloralose is useful for providing long-lasting light anaesthesia for procedures involving no painful surgical interference. A more potent but short-acting anaesthetic can be administered to produce a depth of anaesthesia sufficient to allow surgical procedures to be undertaken, following which unconsciousness can be maintained with alpha-chloralose. Recovery is prolonged and associated with involuntary excitement, and so alpha-chloralose is best used for non-recovery studies. A more detailed discussion of this anaesthetic can be found in Chapter 5.

v. *Urethane*

Desirable effects. Urethane produces long-lasting (6–10 hours) anaesthesia, with minimal cardiovascular and respiratory system depression (Buelke-Sam *et al.*, 1978; Field *at al.*, 1993).

Undesirable effects. Urethane is carcinogenic (Field and Lang, 1988).

Special comments. Urethane resembles chloralose in producing long-lasting stable anaesthesia but, unlike chloralose, the degree of analgesia produced is sufficient to allow surgical procedures to be undertaken in small rodents. It is a useful agent for long-term anaesthesia (see Chapter 5), but it is also a carcinogen and so its use should be avoided whenever possible. If it is necessary to use urethane, precautions appropriate to the handling of a known carcinogen should be adopted. Animals should not be allowed to recover after being anaesthetized with urethane. A more detailed discussion of this anaesthetic can be found in Chapter 5.

II. LOCAL AND REGIONAL ANAESTHESIA

Local anaesthetic agents act directly on nervous tissue to block the conduction of nerve impulses. For example, they can be applied to the surface of the cornea and conjunctiva to produce local anaesthesia of that part of the eye, or they can be used to anaesthetize mucous membranes to ease the passage of catheters or an endotracheal tube.

Local anaesthetics (e.g. lignocaine) can also be injected into tissues to provide a localized area of anaesthesia. Infiltration of the skin and underlying connective tissue will usually provide sufficient anaesthesia to allow the operator to suture minor wounds or to take a biopsy of skin. Infiltration of more extensive areas and the different tissue planes can be used to provide sufficient anaesthesia for surgical procedures such as laparotomy. When injecting local anaesthetics for this purpose, a fine (< 26 gauge), long needle should be used to minimize the discomfort associated with the injection. The syringes used in human dentistry, which are loaded with a cartridge of local anaesthetic, are ideal for this procedure. Care must be taken to infiltrate all of the tissue planes that will be involved in the surgical procedure. If re-insertion of the needle is necessary, this should be done through a previously anaesthetized area, so that discomfort to the animal is minimized. Considerable experience is necessary to ensure that compete blockage of the nerve supply to the surgical field is achieved and expert practical advice should be obtained before carrying out this technique.

If the nerve supply to the operative site is well defined, regional anaesthesia can be produced by infiltration of local anaesthetic around the major sensory nerves. This may involve a single nerve or blockade of several nerves, e.g. in producing a paravertebral block by infiltration of the lumbar spinal nerves as they emerge from the vertebral column, so desensitizing the abdomen.

More extensive effects of local anaesthetic drugs can be produced by injecting them into the spinal canal. The site of injection may be into the fat-filled space between the dura mater and the wall of the vertebral canal (epidural anaesthesia), or directly into the cerebrospinal fluid (subarachnoid or spinal anaesthesia). The techniques for injection have been described both in large animals, such as the cow, sheep and dogs — see Lumb and Jones (1984) for a review — and in laboratory species such as the rabbit (Kero *et al.*, 1981; Hughes *et al.*, 1993) and guinea pig

(Thomasson *et al.*, 1974). When attempting to become proficient in the technique, it is advisable to practise injecting a dye, such as methylene blue, into the spinal canal of a recently killed animal.

A major problem associated with the use of local anaesthetic techniques in laboratory species is that it is often difficult to provide humane, stress-free restraint of the animal during the surgical procedure. It is possible, however, to use the effective surgical anaesthesia which can be produced by these techniques together with low doses of hypnotics or anaesthetics, to provide effective restraint. With some animals, the use of a tranquillizer or sedative, together with the careful attention of an expert handler, may provide sufficient restraint to enable local anaesthesia to be used safely and humanely. When contemplating using local anaesthetic techniques, the likely behaviour of the animal, the type of surgical procedure involved and the expertise of the operator and his or her assistants should be carefully considered.

III. SELECTION OF ANAESTHETIC AGENT: SCIENTIFIC AND WELFARE CONSIDERATIONS

Selection of a particular anaesthetic agent or anaesthetic regimen will depend upon a variety of factors. Somc of these will relate directly to the anaesthetic agent and its potential interactions with the research protocol, others to its ability to produce the required depth of anaesthesia. A further series of factors relate to the practicalities of cost, availability of equipment and the expertise of personnel in the research unit. These various considerations are discussed in more detail below.

Whichever method is chosen, it is important to keep in mind that two primary aims of anaesthesia are to prevent pain and provide humane restraint. The anaesthetic method itself should therefore be one that causes a minimum of distress to the animal. For example, the use of inhalational agents may involve exposure to irritant vapour (e.g. ether, see above), and the restraint required for induction using a face-mask may be stressful. Similarly, restraint for administration of injectable agents can cause distress to the animal, as can the pain associated with injection of certain anaesthetics, and the longer-term consequences of myositis following intramuscular injection of irritant agents (Gaertner *et al.*, 1987; Smiler *et al.*, 1990; Beyers *et al.*, 1991). Other potential problems associated with the use of chloral hydrate and tribromoethanol have been discussed earlier in this chapter.

Intravenous administration usually results in smooth and rapid induction of anaesthesia, provided that the animal is restrained effectively, and that the injection is carried out with the required degree of skill. Local adverse reactions can result from inadvertent perivascular administration (e.g. of thiopentone). Consideration must be given to ways of minimizing any fear or distress associated with handling or physical restraint and movement of the animal from its holding room to the operating theatre or laboratory.

Selecting a method of anaesthesia that is least likely to interfere with a particular research protocol is perhaps the most difficult task. The major pharmacological and physiological effects of the various anaesthetic agents should be reviewed, and this can at least minimize the interactions between the regimen and the research protocol. It is important to appreciate that a superficial consideration of the compound's effects may be insufficient. For example, if one concern is to maintain systemic blood pressure within the range found in conscious animals then, in the rat, pentobarbitone might appear preferable to fentanyl/fluanisone/midazolam in some strains of animal. However, the apparently normal blood pressure is maintained by peripheral vasoconstriction, and cardiac output is markedly depressed (Skolleborg *et al.*, 1990). Animals anaesthetized with fentanyl/fluanisone/midazolam have lower systemic blood pressure, but elevated cardiac output. So it is important to decide which is more important to a particular study — blood pressure or cardiac output. Anaesthetics may sustain blood pressure, but only because of their stimulatory effects on the sympathetic nervous system, so animals may have elevated plasma catecholamine concentrations (Carruba *et al.*, 1987). This information can only be gained by a careful search of the relevant literature. It is important not to assume that such an assessment has been carried out by research workers whose publications include details of the anaesthetic regimen used. Simply adopting the method of anaesthesia described in publications dealing with the particular animal model of interest will not necessarily ensure that an appropriate technique is used.

Having suggested that an assessment of anaesthetic–animal model interactions should be made, it is important also to place such interactions in the context of the overall response to anaesthesia. There is little point in carefully selecting an anaesthetic, then allowing the animal to become hypothermic, hypoxic and hypercapnic because of poor anaesthetic management. These common problems can have wide-ranging effects on the animal's body systems, so attention to good anaesthetic management, described in Chapter 4, is of considerable importance. A second area to consider is the animal's response to surgery. Surgical procedures produce a stress response whose size is related to the severity of the operative procedure. In mammals, this response consists of a mobilization of energy reserves, such as glucose, to enable the animal to survive injury. Although this response has clear evolutionary advantages, it is considered by many to be undesirable in humans and animals who are receiving a high level of intra-operative and post-operative care (Hall, 1985; Salo, 1988). It is also often undesirable because of the potential effects on particular research protocols.

A number of related endocrine responses occur, with elevation in plasma levels of catecholamines, corticosterone or cortisone, growth hormone, vasopressin, renin, aldosterone and prolactin, and a reduction in follicle stimulating hormone, luteinizing hormone and testosterone. Initially, insu-

TABLE 3.6
Checklist of criteria for selection of an anaesthetic regimen for laboratory animals

	Possible anaesthetic agents		
	Anaesthetic 1	Anaesthetic 2	Anaesthetic 3
Is the depth and duration of anaesthesia appropriate?			
Is an appropriate degree of analgesia produced?			
Is the quality of anaesthesia satisfactory?			
Is it easy to assess changes in depth of anaesthesia?			
Is the regimen suitable for the particular species or strain?			
Are there any specific interactions with the experiment?			
Are there are legal or regulatory requirements (e.g. control of narcotics)?			
Is the regimen easy to use?			
Is it reliable and reproducible?			
Is it reversible?			
Is the operator familiar with the regimen?			
What is the cost of the regimen?			
Are all the agents readily available?			

Adapted from Morris *et al.* (1995).

lin concentrations decrease and those of glucagon increase, but later insulin concentrations rise. These hormonal responses produce an increase in glycogenolysis and lipolysis, and result in hyperglycaemia. The duration of the hyperglycaemic response varies, but following major surgery the response may persist for 4–6 hours. More prolonged changes in protein metabolism occur, leading to negative nitrogen balance lasting several days (Hoover-Plow and Clifford, 1978). Even minor surgical procedures can produce relatively prolonged effects. For example, blood vessel cannulation in rats produces an elevation in corticosterone levels for several days (Fagin and Dallman, 1983), and more subtle disruptions of circadian rhythmicity of hormonal secretions can persist for similar periods (Desjardins, 1981)

Research workers are often reluctant to refine their anaesthetic methodology because it is thought that the anaesthetics used in the new regimen may affect their animal model in the post-surgical period. In some instances there will be a sound scientific basis for this opinion, based on a critical review of the relevant literature. In other circumstances, the effects of anaesthesia may be relatively unimportant when compared with the effects of surgical stress. Similar concerns are also expressed about the use of post-operative analgesics, and once again the side-effects of any analgesics used should be considered alongside the other effects of surgery and anaesthesia. Clearly, it is logical to consider all of the factors that may interact with a particular study, and develop an anaesthetic and surgical procedure which is both humane and provides the minimum of interference with the overall aims of the research project.

Many of the significant considerations involved in selecting an anaesthetic agent were reviewed by a small working group (Morris *et al.*, 1995) and this group's suggestion of tabulating relevant factors to simplify the selection process may be found helpful (Table 3.6).

4

Anaesthetic Management

1. PRE-OPERATIVE PREPARATIONS

Following induction of anaesthesia with one or more of the drugs described in Chapter 3, the animal should be placed in a suitable position to enable the required surgical procedures to be carried out. A compromise must usually be made between a position considered ideal by the surgeon and one that avoids compromising the function of any of the animal's body systems. In particular, care must be taken to ensure that the head and neck remain extended, so that the tongue or soft palate does not obstruct the larynx. The limbs should not be tied out in such a way that thoracic respiratory movements are impeded; and care must be taken during surgery that undue pressure is not placed on the chest wall or abdomen. Over-enthusiastic use of retractors and the use of the thorax as an arm-rest during surgery are commonly observed and must be discouraged! In smaller species, the use of elastic bands to position the animal can often lead to excessive extension of the limbs and consequent interference with respiratory movements. Similarly, elastic bands placed around the abdomen can interfere both with diaphragmatic movements and the venous return from the hindquarters and abdominal viscera. This, and similar techniques that aim at producing an immobile animal, are rarely necessary and should be discouraged.

If it becomes necessary to tie out limbs they should not be pulled into full extension and the anchoring bandages should be tied loosely. This is particularly important during prolonged anaesthesia, when constricting ties around the limbs can lead to tissue damage and peripheral limb oedema, which is likely to cause considerable discomfort to the animal in the post-operative period. If an endotracheal tube is used, it should be tied firmly to the animal's jaw. It is often helpful to tape the anaesthetic circuit to the operating table, so that it cannot drag on the endotracheal tube and dislodge it. The risk of inadvertent disconnection is reduced by using lightweight disposable circuits (see Chapter 3). Particular care must be taken if an animal requires repositioning during an operative procedure. Turning the animal may result in kinking of the endotracheal tube and it is usually preferable temporarily to disconnect the animal from the breathing circuit while moving it.

During anaesthesia, the protective reflexes that prevent damage to the eye are often lost, so the cornea is susceptible to desiccation and damage. To reduce this danger, the eyelids should be taped closed with a small piece of adhesive dressing, or filled with an ophthalmic ointment.

II. MONITORING ANAESTHESIA

Following the administration of an anaesthetic, it is essential to assess that the required depth of anaesthesia has been achieved. It is also helpful to monitor the vital signs of the patient and the function of any anaesthetic apparatus that is in use.

A. Assessment of depth of anaesthesia

Ensuring that an animal is maintained at the correct depth of anaesthesia requires the development of some degree of clinical skill by the anaesthetist. With the wide range of anaesthetic drugs currently available, the simple classical approach of dividing anaesthesia into a series of levels and planes (Lumb and Wynn Jones, 1973) is now of limited use. This classification of anaesthetic depth relied heavily on the assessment of cardiovascular and respiratory function and was developed in humans to enable the safe use of a single volatile agent (ether) to produce deep surgical anaesthesia (Bendixen, 1984). The widespread variation in response to anaesthesia in different animal species and the use of several drugs in combination makes such a scheme virtually unworkable in modern laboratory animal anaesthetic practice. Although there is little scope for the production of a general classification of depth of anaesthesia, a range of clinical observations can be made to aid in the assessment of the depth of anaesthesia. Following administration of a volatile anaesthetic agent or the intraperitoneal injection of a drug such as pentobarbitone, most animals will become ataxic, lose their righting reflex and eventually remain immobile. At this depth of anaesthesia they can easily be roused by painful stimuli; so anaesthesia must be allowed to deepen until such responses to pain are absent. This sequence of events will not be seen if induction is carried out by the rapid intravenous injection of an anaesthetic such as alphaxalone/alphadolone or propofol. More sophisticated techniques of assessment of depth of anaesthesia have been developed in humans, for example measurement of the electroencephalograph and of sensory or somatic evoked potentials. These have not yet been widely applied in animals (Whelan and Flecknell, 1992).

1. Responses to painful stimuli

Because the reason for inducing anaesthesia is often to block the perception of pain, the response to a painful stimulus is an essential part of the assessment of the depth of anaesthesia. In most species, the pedal withdrawal reflex should be assessed. One limb should be extended and the web of skin between the toes pinched between the anaesthetist's fingernails. If the limb is withdrawn, muscles twitch, or the animal cries out, the depth of anaesthesia is insufficient to allow surgical procedures to be carried out. In small rodents it may be difficult to pinch the toes and pinching the tail provides a convenient alternative stimulus. Besides using the limb withdrawal response, the reaction to pinching an ear can be observed in rabbits or guinea pigs. At light levels of anaesthesia, the animal responds to ear pinching by shaking its head and at very light levels by vocalizing. The loss of a response to painful stimuli does not occur uniformly in all body areas. Occasionally, it may be possible to begin to perform a laparotomy without eliciting either any movements or any autonomic responses, such as an increase in heart rate, in an animal which still shows a limb withdrawal reflex. Further surgical stimulation, such as cutting or clamping the abdominal muscles or handling the abdominal viscera, will often produce reactions indicating an inadequate depth of anaesthesia. Consequently, it is recommended that absence of a limb withdrawal response is used as an indication of onset of surgical anaesthesia and that less sensitive indicators such as the response to a pin-prick on the abdominal skin are not relied upon.

2. Alterations in eye reflexes

Examination of eye reflexes and of the position of the eyeball are of limited use in small laboratory species. In larger species, such as the dog, cat, pig, sheep and primates, the palpebral reflex (blinking when the edge of the eyelids is lightly touched) is lost during the onset of light surgical anaesthesia with barbiturates, volatile anaesthetics and some other drugs. Use of ketamine will cause the loss of this reflex at lighter levels of anaesthesia, and the use of neuroleptanalgesic combinations has unpredictable effects on the reflex. The palpebral reflex is difficult to assess in small rodents, and in rabbits it may not be lost until dangerously deep levels of anaesthesia have been attained. The position of the eyeball can also be of use once experience has been gained with the species of animal and with the particular anaesthetic technique which is to be used. The position of the eyeball, the degree of pupillary dilation and occurrence of side-to-side movement (nystagmus) cannot generally be relied upon as indicators of the depth of anaesthesia and should always be combined with observation of other clinical signs.

3. *Alterations in cardiovascular and respiratory functions*

Most anaesthetics cause a dose-dependent depression of the cardiovascular and respiratory systems. The way in which this depression is manifested can vary considerably with different anaesthetics. Respiratory rate may decrease, or may increase in response to a fall in the depth of respiration. Conversely, the depth of respiration may increase and respiratory rate decrease. Cardiovascular system depression usually results in a fall in systemic blood pressure, but this may be associated with either a fall or a rise in heart rate.

Given these wide variations in response, it is dangerous to generalize about the effects of anaesthetics on these body systems. Once again, experience with a particular anaesthetic regimen and particular animal species will be needed before these changes can be interpreted with any confidence. Details of methods of monitoring the cardiovascular and respiratory system are given below.

B. Assessment of patient well-being

All anaesthetic agents produce a reversible depression of many central nervous system activities and, on occasion, the degree of depression will be excessive and the animal will die. It is important that a certain percentage of anaesthetic deaths does not become accepted as an unavoidable consequence of anaesthesia. The death of an animal during anaesthesia should be a stimulus to review the entire process of animal selection and of pre-operative and intra-operative care. Anaesthetic mortality rate can usually be reduced by observation of the precautions described in Chapter 1. In addition, careful assessment of the physiological state of the animal during the period of anaesthesia can result in a dramatic improvement in recovery rates. Monitoring of an anaesthetized animal does not necessarily involve the use of complex electronic apparatus; although (as discussed later) such equipment can prove extremely valuable. Even when using sophisticated devices, basic clinial observations such as noting the colour of the mucous membranes, the pattern and rate of respiration and the rate and quality of the pulse, are of fundamental importance. These simple clinical observations can be carried out by anyone and will often indicate a deterioration in the animal's condition before the deterioration becomes irreversible.

Although simple clinical observation by the anaesthetist should never be neglected, when the roles of anaesthetist and surgeon or anaesthetist and theatre nurse are combined, uninterrupted or even regular observation is often impossible. In addition, fatigue during prolonged procedures may lead to human error, so the use of electronic equipment to provide continuous monitoring of physiological variables can be invaluable. Certain variables can only be measured directly by using electronic equipment, and, when anaesthetizing animals for prolonged periods or during complex or high-risk procedures, the additional information provided by such apparatus can greatly assist management of the anaesthetic. A further advantage of electronic monitoring equipment is that it usually enables acceptable limits for each monitored variable to be set at the start of the period of anaesthesia. An audible or visible alert is triggered when these pre-set limits are exceeded. As mentioned earlier, the degree of monitoring required will depend upon the nature and duration of the surgical procedure.

Whatever monitoring is to be undertaken, it is of fundamental importance to make some record of the information obtained. In almost every case problems develop gradually, rather than occur as sudden catastrophes. If the observations made are recorded, preferably as simple graphs, adverse trends are easily detected and appropriate corrective action can be taken. A second advantage of such a record is that it enables a series of anaesthetics to be critically reviewed, so that techniques can be evaluated and improved.

As discussed earlier, it is only through practical experience that the ability to assess the significance of changes of physiological variables during the administration of a particular anaesthetic can be developed. Although the production of written records may seem unduly time-consuming, they provide an invaluable source of reference both for the current anaesthetist and for less experienced staff who may be required to undertake the procedure in the future.

C. Respiratory system

1. Clinical observations

A number of observations can assist in detecting a deterioration in respiratory function. The rate, depth and pattern of respiration can be assessed by observation of the animal's chest wall, or of the anaesthetic circuit's reservoir bag if one is present. In larger animals, an oesophageal stethoscope can be used to monitor breath sounds as well as heart sounds.

a. Respiratory monitors. The respiratory rate can be conveniently monitored with an electronic monitor. The most inexpensive of these use a thermistor, which is either mounted in the anaesthetic circuit in the endotracheal tube or face-mask connector, or placed close to the animal's external nares. The response of some of these devices is sufficiently sensitive to enable monitoring of respiration in animals weighing as little as 300 g. An alternative technique for monitoring respiratory rate relies upon movements of the chest wall to trigger a pressure sensor (Graseby Medical, Appendix 7). These devices can be used with most animals weighing over 1000 g. When buying a respiratory monitor, ensure that its sensitivity is sufficient to function reliably with the species you intend to anaesthetize and that an alarm can be set to detect apnoea. If buying a more sophisticated instrument that allows upper and lower limits for respiratory rate to be set, make sure these are wide enough for the full range of species that will be monitored.

b. Measurement of tidal and minute volume. Although apnoea alarms will indicate changes in the rate of respiration, it is also useful to assess the depth of respiration by measuring tidal volume. Both tidal and minute volumes (the volume of gas breathed in 1 minute) can be measured using a Wright's respirometer. The standard model of this instrument will measure tidal volumes down to 200 ml. A paediatric version is also available which has a range of 10–250 ml and is appropriate for use in animals ranging in size from large guinea pigs to medium-sized dogs.

Respirometers are used primarily to perform intermittent measurements of tidal and minute volume, and are most widely used for assessing that mechanical ventilators are delivering an appropriate volume of gas. The relatively large dead space of the instrument makes it impracticable for permanent placement in the circuit with a small animal, and in all sizes of patients build-up of water vapour can cause the instrument to fail.

c. Assessment of lung gas exchange. Although measurements of the mechanical aspects of respiration usually provide a reasonable indication of respiratory function, some attempt must also be made to assess the adequacy of lung gas exchange. This can be judged clinically simply by observing the colour of the visible mucous membranes and the colour of any blood which is shed at the site of surgery. Although such simple clinical monitoring will show the onset of severe hypoxia, it gives no indication of blood carbon dioxide content. A more sensitive measure of blood oxygen saturation can be obtained using a pulse oximeter (see below).

i. Pulse oximetry. Pulse oximeters measure the percentage saturation of arterial blood, by detecting changes in the absorption of light across the tissues. A variety of probes of different shapes and sizes are available, the majority designed for human use. Both re-usable and disposable probes can be obtained. Besides measuring the saturation of the haemoglobin with oxygen ($Sa\text{O}_2$), the instrument measures the pulsatile nature of the signal, and from this calculates the heart rate. Although the absorption spectra of haemoglobins vary between species, they are sufficiently similar to allow instruments designed for human use to be used successfully in most mammals (Decker *et al.*, 1989; Erhardt *et al.*, 1990; Vegfors *et al.*, 1991; Allen, 1992; Jacobson *et al.*, 1992). Attaching a pulse oximeter gives three useful pieces of information. The degree of saturation of haemoglobin allows detection of hypoxia due to respiratory depression, airway obstruction or anaesthetic equipment failure. The heart rate reading is useful in detecting changes in rate associated with cardiovascular system depression, or tachycardias caused, for example, by carrying out surgical procedures at an inadequate anaesthetic depth. Finally, the strength of the pulsatile signal, usually displayed as a bar graph or as a wave-form, provides some indication of the flow of blood through the tissues. This is often more informative than a simple indication of heart rate, since it reflects the mechanical action of the heart.

Pulse oximeters have been shown to be reasonably accurate at normal oxygen saturation levels (80–99%), but to become increasingly inaccurate as saturation falls. They should therefore only be used to provide a general indication of the adequacy of tissue oxygenation, and cannot be relied upon to accurately record low saturations. Nevertheless, these devices are considerably more reliable than simple clinical assessment, and since development of low saturation requires immediate corrective action, their relative inaccuracy in this range is rarely of clinical importance.

Pulse oximeters are sensitive to movement artefacts, and this can cause difficulty if these instruments are used in the later stages of recovery from anaesthesia. They will fail to provide a signal if the pulsatile blood flow through the tissues falls, as occurs during shock. In very small animals, the low volume of tissue available for monitoring limits the reliability of currently available instruments. Several manufacturers' instruments can be used successfully in animals weighing more than 200 g, although the upper heart rates displayed, and the corresponding high heart rate alarm, are limited to 250 beats per minute (bpm.). Instruments specifically designed for animal use are becoming available (Vet/Ox, Sensor Devices

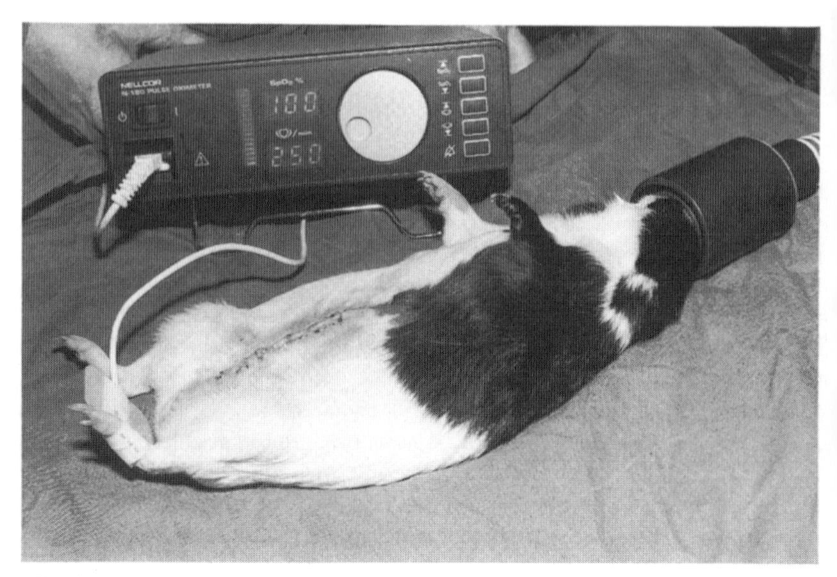

Fig. 4.1. Pulse oximetry in the guinea pig. A disposable paediatric probe is placed across the animal's hind foot.

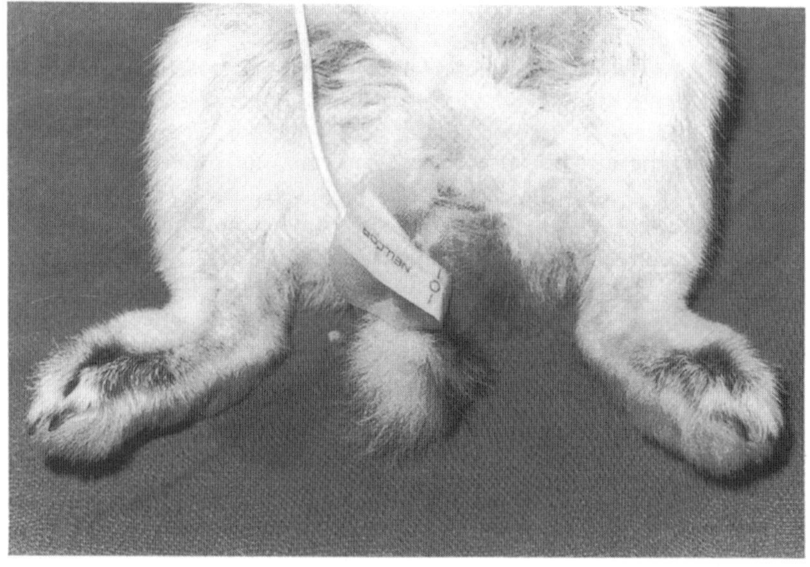

Fig. 4.2. Pulse oximetry in the rabbit. A disposable paediatric probe is placed across the animal's tail.

Inc., Appendix 7), with upper rate limits of 350 bpm. Suitable sites for probe placement include the tongue, ears, tail, nail-bed, and across the foot pad in rats and guinea pigs (Figures 4.1 and 4.2).

ii. End-tidal carbon dioxide. An indication of carbon dioxide exchange can be obtained by monitoring the concentration of carbon dioxide present in the exhaled gas using a capnograph. The maximum concentration detected, the end-tidal concentration, reflects the concentration of carbon dioxide present in alveolar gas. Considerable additional information can be obtained from the wave-form which shows the changing concentration of carbon dioxide during the respiratory cycle. A capnograph can alert the anaesthetist to abnormal build-up of carbon dioxide caused by respiratory failure, and also to rebreathing of exhaled gas caused by inadequate fresh gas flows. Capnographs are designed either to sample expired gases from a tube placed close to the endotracheal tube (side-stream systems) or to have the carbon dioxide sensor placed directly in the anaesthetic circuit (mainstream sampler). Although mainstream samplers have some advantages in respect of sensitivity and speed of response, they have the disadvantage of introducing additional dead space in the circuit. Side-stream samplers are generally satisfactory for most species, but it is important to establish the instrument's sampling rate. Most capnographs sample 150–200 ml of gas per minute, but many have a paediatric setting of around 50 ml min^{-1}. Since the minute volume of a 200 g rat will be approximately 120–200 ml min^{-1}, even sampling at the lower rate of 50 ml min^{-1} can lead to dilution of the expired gas sampled with gas from the anaesthetic circuit, and an underestimate of the end-tidal carbon dioxide concentration. Provided that the gas flow in the circuit is not altered, capnograph readings will indicate trends during the anaesthetic procedure in these small animals, and so are useful, particularly in animals maintained using a mechanical ventilator. If blood gas analysis is carried out at the start and end of the procedure, then the capnograph readings can be related to arterial P_{CO_2} values, and the constant read-out from the capnograph will provide reassurance that the carbon dioxide tension is constant.

iii. Blood gas analysis. The most satisfactory method for measuring the adequacy of lung gas exchange is to obtain arterial blood samples and carry out blood gas analysis. Blood gas analysers will measure the partial pressure of oxygen and carbon dioxide and the pH of the blood and, in addition, will calculate the blood bicarbonate concentration and the base excess. Analysers designed for human paediatric use require sample volumes as low as 0.1 ml, so their use in monitoring blood gases in small animals becomes practicable. A major problem will often be the difficulty in obtaining arterial blood samples in smaller animals. It is also important to appreciate that the temperature of the patient must be recorded, since the instrument applies a correction to its measurements based on this. In view of the common occurrence of hypothermia in small animals, this can be a significant source of error. Instruments designed for human use carry out their calculations based on data from human haemoglobin. Results are generally applicable to animals, but for greater accuracy instruments are available that allow data on animal haemoglobins to be used.

A reasonably close approximation of arterial carbon dioxide and oxygen concentration may be obtained non-invasively and continuously by using transcutaneous oxygen and carbon dioxide monitors. These instruments have been little used in animal anaesthesia, but limited experience suggests that they can be used successfully in sheep, lambs and rabbits.

D. Cardiovascular system

1. Clinical observations

The rate, rhythm and quality of the peripheral pulse can be assessed in rabbits, cats and larger animal species. The femoral artery is easily palpable, but if the animal is covered with sterile surgical drapes, both this and other pulse points may be inaccessible. In the dog, the sublingual artery and the digital artery can be palpated, but some practice is needed before these pulse points can be used with confidence. The assessment of the quality of the pulse will give a rough indication of the adequacy of systemic arterial pressure. Some indication of the adequacy of tissue perfusion can be gained by observing the capillary refill time in the visible mucous membranes. The gums are usually the most accessible site and the refill of the capillaries following blanching by digital pressure can be observed in most larger species.

The heart sounds and heart rate can be assessed by use of a stethoscope positioned on the chest wall or, in dogs and larger animals, by means of an oesophageal stethoscope.

a. Electrocardiography. The electrical activity of the heart can be monitored by an electrocardiogram (ECG). Instruments designed for use in humans are normally acceptable for monitoring animals, but the maximum heart rate that can be displayed is usually 200 or 250 beats per minute. Small rodents and rabbits frequently have heart rates in excess of 250 bpm. and this may limit the usefulness of some of these monitors.

Purpose-designed instruments for animal use are now becoming available (Silogic, Appendix 7), which enable low-voltage ECG signals and rapid heart rates to be detected. Electrocardiograph electrodes designed to stick on the skin can be used successfully in larger animals, provided that any hair in the area of electrode placement is carefully removed. Human paediatric electrodes are suitable for cats, rabbits and small primates, but needle electrodes are usually required for small rodents. Electrode placement on the left and right forelimb and right hindlimb will provide a standard ECG trace, but the signal amplitude from small animals may be insufficient to produce an adequate display on some monitors. Whenever possible, arrange to have a demonstration of an electrocardiogram on the species of interest before purchasing the instrument.

Electrocardiographs sold for medical and veterinary use are designed either for diagnostic electrocardiography or for patient monitoring. In general, the diagnostic instruments produce output to a pen recorder and do not have facilities for triggering alarms. Patient monitors have a visual display unit (VDU), usually with heart rate displayed as well as the ECG trace. Upper and lower rate limits can usually be set, although the restricted range of these settings in some instruments may limit their use in some animals. It is important to appreciate that the ECG indicates only the electrical activity of the heart and does not indicate adequate circulatory function. It is possible to have a cardiac output of zero and a normal ECG!

Automated heart rate meters can be obtained that measure electrical activity of the heart but display only the rate. A heart rate can also be obtained from a pulse oximeter (see above), or by using a Doppler flow probe positioned over a suitable artery.

b. Blood pressure

i. Systemic arterial pressure. Direct or indirect recording of systemic arterial blood pressure is possible in most laboratory animal species. Direct measurements are invasive, requiring arterial cannulation, but they can be applied easily to most species. Cannulation can be done either following surgical exposure of a suitable artery or by percutaneous puncture using a catheter and introducer. The femoral artery can be cannulated in this way in dogs, pigs, sheep and larger primates. Percutaneous cannulation of the femoral artery in the cat and rabbit requires considerable technical skill. In rabbits and sheep the central ear artery provides a convenient vessel for percutaneous catheterization.

Invasive blood pressure monitoring has the advantage of providing a rapid response to changes in pressure and in recording accurately over a wide range of blood pressures. Blood pressure can also be monitored non-invasively using a sphygmomanometer. Instruments designed specifically for use in animals are now available, and are preferable because use of an appropriately sized cuff for occlusion of the artery is essential for accurate measurement (Kittleson and Olivier, 1983). A Doppler probe to detect arterial blood flow, coupled with an inflatable cuff and a pressure sensor, can be used to measure arterial blood pressure in a range of animal species. These instruments are available commercially and can be used to measure arterial pressure in the caudal artery of rats (Harvard Apparatus Ltd, Appendix 7).

The main disadvantage of non-invasive monitoring is the intermittent nature of the information obtained. The most widely used automated instruments, which use an oscillometric technique to detect the arterial pressure changes, take readings at a minimum interval of 1 minute. During periods of cardiovascular instability, this interval may be considered unacceptable. A second problem is that when blood pressure falls, the instrument may fail to detect the weaker signals.

Pulse oximetry, described above, provides a measure of heart rate and gives a crude but effective indication of the pulsatile flow in the tissues.

ii. Central venous pressure. Central venous pressure can be measured by inserting a cannula into the jugular vein and advancing it so that its tip lies in the anterior vena cava. The catheter can be introduced percutaneously or following surgical exposure of the vein. The simplest method of recording central venous pressure is to connect the cannula to a water manometer which has had its baseline (zero) reading set at the estimated level of the animal's right atrium. Water manometer systems are generally unsatisfactory in smaller animals, such as rodents and rabbits, and it is preferable to use an electronic pressure transducer.

Pressure transducers for arterial and venous pressure are expensive items of equipment, but a range of disposable transducers for human use are now available. These are much cheaper to buy, and provided that absolute asepsis is not required they can be re-used successfully for a considerable period.

E. Body temperature

Body temperature is one of the easiest physiological variables to monitor during anaesthesia. Rectal temperature can be monitored simply by using a clinical thermometer, but this will require repeated adjustment and replacement of the instrument in the rectum to record the changes in body temperature that may occur during anaesthesia. A second disadvantage is that the lowest temperature measurable may be only 35°C. The body temperature of small animals can easily fall below this value, and the onset of hypothermia can be overlooked. It is much more satisfactory to purchase one of the wide range of electronic thermometers that can provide continuous display of body temperature. The rectum is often the most convenient site for placing a temperature probe, but deep body or core temperature will often be underestimated, and if the probe is positioned in the middle of a mass of faeces its response time to changes in temperature will be slow. For these reasons it is often preferable to use a probe placed in the oesophagus, but it must be located in the lower part of the oesophagus to avoid the cooling effects of respiratory gases in the upper airway.

Measurement of skin surface temperature is also valuable and this can conveniently be carried out by taping a temperature probe between the animal's digits. In a healthy anaesthetized animal the temperature difference between the core and the periphery rarely exceeds 2–3°C. An increase in this temperature gradient indicates that peripheral vasoconstriction is occurring and the possible causes of this should be investigated.

If routine temperature monitoring of a range of different species is to be undertaken, it is worth purchasing a more sophisticated electronic thermometer that allows the simultaneous use of temperature probes of different designs. Suitably sized probes for rectal or oesophageal placement in mice, rats, rabbits and larger animals are available, together with skin surface temperature probes, needle probes and other special purpose probes. Thermometers for measuring temperature at the tympanic membrane in humans, which have an extremely rapid response time, may be used in animals but their accuracy and reliability vary considerably when used in laboratory species. If use of one of these devices is contemplated, it should be carefully validated using a conventional electronic probe before reliance is placed on the data obtained.

F. Anaesthetic equipment function

Before administering an anaesthetic agent, it is important to check that all the equipment to be used is functioning correctly (see Chapter 1). Even if all the equipment is functioning correctly at the start of the anaesthetic procedure it is important to monitor its continued normal function.

1. Circuit disconnection

The risk of inadvertent disconnection of the animal, the anaesthetic breathing circuit and the anaesthetic machine can be reduced by using safe-lock connectors. The most frequent point of disconnection is at the junction of the circuit and the endotracheal tube. It is possible to position a thermistor-type apnoea alarm in the breathing circuit and this can provide an alert if disconnection occurs. When anaesthetizing larger animals, use can be made of pressure monitoring in the breathing circuit, which will detect both low pressure due to disconnection and high pressure caused (for example) by a malfunctioning expiratory valve. An oxygen analyser, positioned within the fresh gas flow of the circuit, will detect disconnection of the circuit from the anaesthetic machine and also any failure of the oxygen supply. Some machines are fitted with an audible alarm that is activated if the oxygen pressure falls below a lower limit.

2. *Infusion pumps*

If anaesthesia is being administered by continuous intravenous infusion, it is useful to have a warning system that will detect pump failure or exhaustion of the infusate. This is particularly important if neuromuscular blocking agents are used, which would prevent any spontaneous movements of the animal that might occur as the depth of anaesthesia was reduced. Older style infusion pumps generally have no warning devices; so it remains the anaesthetist's responsibility continuously to monitor their function. The more recently available microprocessor-controlled infusion pumps may be fitted with a variety of devices to alert the anaesthetist to a malfunction, but even these should be regularly inspected throughout the operative period. The main features to consider when purchasing such devices are listed in Chapter 5.

III. ANAESTHETIC PROBLEMS AND EMERGENCIES

Monitoring the state of the animal during anaesthesia will give early warning of impending problems and emergencies, so that corrective action can be taken. There is little purpose in adopting the monitoring procedures described earlier unless the information obtained is of value, and influences the course of action taken should problems arise. In clinical anaesthesia, the successful resuscitation of the patient is of paramount importance, but in a research setting, additional factors must be considered. An animal that has developed problems during anaesthesia, for example severe respiratory depression, may no longer be a suitable animal model for some types of study. A second consideration is that extensive emergency therapy may result in additional pain and distress to the animal concerned. These factors must be considered, preferably before administering an anaesthetic, so that an appropriate course of action in the event of emergencies can be planned.

Given these constraints, the following section outlines the major indications of impending problems and suggests appropriate corrective measures.

A. Respiratory system

1. Signs of impending failure

a. Respiratory rate. Respiratory rate should be recorded before anaesthesia, so that any depression in rate can be assessed. If the animal is calm and relaxed, then the assessment will be reasonably accurate, but many rodents and rabbits will have a marked increased respiratory rate in the immediate pre-anaesthetic period, caused by fear and apprehension. In these circumstances, all that can be done is to make an estimation of the normal respiratory rate based on published data (Appendix 1). During anaesthesia in rodents and rabbits, as a general guide, a fall in the respiratory rate to less than 40% of the pre-anaesthetic rate indicates impending respiratory failure. Changes in respiratory rate with varying depths of anaesthesia differ depending upon the agent used, so considerable experience is needed to assess their significance across a wide range of techniques and species. Nevertheless, when using a single anaesthetic in one species, it is relatively easy to develop an appreciation for the effects of increased or decreased depth of anaesthesia. This learning process can be speeded by always taking the opportunity to observe animals that have been killed with an overdose of anaesthetic and, if possible, to use an overdose of the intended anaesthetic regimen as the euthanasia agent.

A rise in respiratory rate may be due to a lightening of the level of anaesthesia, and may also require corrective action. The animal should be carefully assessed for other signs of a reduced depth of anaesthesia (see below), since an increased respiratory rate and depth may also occur if carbon dioxide accumulates in the breathing circuit. This can occur during closed-circuit anaesthesia if the soda-lime carbon dioxide absorber has become depleted, or in any circuit if there is a failure of the fresh gas supply.

b. Tidal volume. A progressive fall in tidal volume frequently indicates impending respiratory failure. As with most other monitored variables, it is important to record the trends that occur during the period of anaesthesia. An apparent sudden failure of respiration is nearly always preceded by a progressive deterioration in tidal and minute volumes.

c. Lung gas exchange

i. Mucous membranes. Any noticeably blue colouration of the visible mucous membranes indicates the onset of severe hypoxia. In most species oxygen saturation may fall below 50% before any evidence of cyanosis is detected. It is therefore important to regard development of cyanosis as an emergency requiring immediate corrective action. The colour change of the mucous membranes indicates only a lack of oxygen, and the membranes may remain a normal pink colour even in an animal with a grossly elevated blood carbon dioxide content.

ii. Pulse oximetry. A more accurate assessment of the degree of oxygenation of the arterial blood can be obtained using a pulse oximeter. In animals breathing room air, saturation is normally 95–98%. Animals breathing oxygen will have a saturation of 100%. Falls of more than 5% should alert the anaesthetist to the onset of mild hypoxia, and a reduction of more than 10% requires immediate corrective action. Readings below 50% indicate severe, life-threatening hypoxia.

iii. End-tidal carbon dioxide. Spontaneously breathing animals have an end-tidal carbon dioxide concentration in the range 4–8% and during artificial ventilation a concentration of 4–5% should be maintained if arterial carbon dioxide tensions are to be kept within the normal physiological range. A gradual rise in end-tidal carbon dioxide concentration indicates progressive hypercapnia, and corrective action should be taken. Increased concentrations can also occur because of failure of the fresh gas supply, exhaustion of the soda lime during closed-circuit anaesthesia, or problems with the anaesthetic circuit. If the capnograph trace fails to return to zero this usually indicates that rebreathing of exhaled gas is occurring, and this can be prevented by increasing the fresh gas flow or reducing anaesthetic circuit dead space. A gradual fall in carbon dioxide concentration can indicate increased ventilation, but may also occur during hypotension and decreased cardiac output. Sudden reductions in end-tidal carbon dioxide concentrations can indicate airway obstruction, disconnection of the animal from the breathing circuit, or cardiac arrest. Assessment of the respiratory rate and tidal volume will help distinguish the likely cause. As experience is gained, considerable information can be obtained from the capnograph wave-form (Cruz *et al.*, 1994).

d. Blood gases. It is important to establish a baseline measurement of blood gas values as soon as possible following the induction of anaesthesia. This may be compared with normal values for the particular species, but

TABLE 4.1
Blood gas values for animals breathing air

	Arterial blood		Venous blood	
P_{CO_2}	3·8–5·3 kPa	(28–40 mmHg)	3·8–5·6 kPa	(28–42 mmHg)
P_{O_2}	11–12·5 kPa	(82–94 mmHg)	5·3–8 kPa	(40–60 mmHg)
pH	7·35–7·45		7·3–7·9	

these are broadly similar for most animals (Table 4.1). A progressive fall in blood oxygen concentration and/or a rise in carbon dioxide concentration, usually accompanied by a fall in pH, indicates inadequate gas exchange. Animals breathing room air (20% oxygen) will normally have an arterial P_{O_2} of 11–12.5 kPa (82–95 mmHg); a fall below 10.5 kPa (80 mmHg) requires corrective action. It is important to note that animals receiving oxygen will normally have much higher arterial partial pressures of oxygen. Values in the range 40–53 kPa (300–400 mmHg) can be anticipated. In these circumstances, a fall in P_{O_2} below 12–15 kPa (90–112 mmHg) in an animal breathing 40–60% oxygen should be considered serious. A rise in P_{CO_2}, from a typical baseline of 5 kPa (37.5 mmHg) to above 6.5 kPa (50 mmHg) indicates mild to moderate hypercapnia. Increases greater than 8 kPa (60 mmHg) indicate severe hypercapnia and consequent respiratory acidosis. Detailed interpretation of blood gases data is complex, but not difficult to master. An excellent source of reference is provided by Martin (1992).

2. *Corrective action*

Impending respiratory failure requires immediate corrective action. The following list provides a quick guide to dealing with the most frequently encountered problems:

- If an anaesthetic circuit and a source of oxygen is in use, quickly check that oxygen is still being supplied.
- Check that the circuit is correctly assembled and still connected to the animal.
- If volatile agents are in use, reduce the concentration to zero and remember to switch off any nitrous oxide and to increase the oxygen flow rate to compensate for the reduced total gas flow.
- If injectable agents are being administered, stop any continuous infusions and consider whether a reversal agent should be administered. For example, if a neuroleptanalgesic anaesthetic combination has been used, respiratory depression can be reversed using a specific antagonist such a nalbuphine or diprenorphine (Revivon, C-Vet) (see Table 7.3). Since

this will reverse the anaesthetic and analgesic actions of the combination, it must not be administered if surgical procedures are still in progress.

- Fill the circuit with oxygen using the emergency oxygen switch on the anaesthetic machine and assist ventilation as described in Chapter 3 for a few respiratory cycles.
- Observe the movement of the chest to ensure that gas is moving in and out of the lungs. If it is not, check the endotracheal tube — this may have become kinked, be pushed too far down the airway, or have become blocked with secretions.
- If excessive secretions are present (indicated by bubbling noises during ventilation), disconnect the circuit and clear the tube using gentle suction. Although a vacuum suction device is very useful, a simple technique is to use an appropriate-sized catheter attached to a 10–50 ml syringe.
- If the animal has not been intubated, check that the head and neck are extended, open the animal's mouth and pull its tongue forward to ensure that it is not obstructing the larynx.
- If oxygen is not being administered but an oxygen supply is available, administer 100% oxygen as soon as possible. If an anaesthetic circuit is in use, continue assisting ventilation; if a circuit is not connected, assist ventilation by manual compression of the thorax. This can be carried out successfully even in small rodents, when compression with the thumb and forefinger can be used to produce some respiratory gas movements.
- Consider other possible causes of depressed respiration. Check on the activities of the surgeon: for example, whether movements of the animal's chest are being restricted, either by using the thorax as a support or by inappropriate positioning of retractors or packs.
- Respiration can be stimulated by the administration of an analeptic such as doxapram. This agent can be used in all species (see Table 7.3), but as its action is of relatively short duration repeated doses may be required every 15–20 minutes.
- If assisted respiration and administration of oxygen have improved respiratory function, observe the animal carefully to check whether any deterioration occurs when these measures are stopped. If respiration appears stable, recommence administering the anaesthetic drugs if these have been stopped and continue to observe the animal carefully. If respiratory function deteriorates, recommence assisted ventilation and preferably connect the animal to a mechanical ventilator. Try to reduce the depth of anaesthesia, but this may be limited by the continuance of any surgical procedures.

B. Cardiovascular system

Most anaesthetic drugs have a depressant effect on cardiovascular function, and overdosage of anaesthetic is probably the most common cause of cardiac failure. Both the heart rate and the force of contraction can be depressed and, in addition, cardiac arrhythmias may occur. These may also be caused by hypoxia and hypercapnia due to respiratory failure. If the circulation is severely depressed and insufficient oxygen is delivered to the body tissues, peripheral circulatory changes may occur which lead to the development of shock. Besides the adverse effects of anaesthesia or respiratory system failure, loss of blood and body fluids may result in a reduction in the circulating blood volume. If blood volume falls excessively, cardiovascular failure and cardiac arrest will occur. Severe hypothermia (body temperature approximately 25°C) will also result in cardiac arrest.

1. Clinical signs of cardiac failure

a. Appearance of mucous membranes. Progressive circulatory failure may be detected by a deterioration in capillary refill time, assessed by blanching the peripheral mucous membranes by digital pressure. Any noticeable delay in refill indicates a reduction in tissue perfusion. A reduction in tissue perfusion may also produce a moderate cyanosis (bluish tinge) of the mucous membranes. Cyanosis is more frequently associated with respiratory failure and, if due to cardiovascular failure alone, it indicates severe circulatory disturbance. Severe circulatory failure due to hypovolaemia will produce a blanching of the visible mucous membranes.

b. Peripheral temperature. Severe circulatory failure is also associated with a fall in peripheral temperature and the animal's limbs will be noticeably cool to the touch. This can be detected more readily by using a temperature probe taped between the animal's digits and comparing the peripheral temperature with rectal temperature. This temperature change develops slowly, so will not be of immediate use if rapid haemorrhage has occurred.

c. Blood pressure. Systematic arterial pressure will fall during the development of cardiac failure. Usually the fall is gradual and regular monitoring of blood pressure will allow corrective action to be taken before severe changes have occurred. If a pulse oximeter is in use, a fall in signal strength or complete loss of the signal may occur because of hypotension. As mentioned above, a fall in end-tidal carbon dioxide tensions can indicate hypotension. It is advisable to maintain mean arterial blood pressure above 60–70 mmHg, to avoid problems caused by poor tissue perfusion. A fall in mean arterial pressure below 45 mmHg can result in a failure of renal blood flow, severe metabolic disturbances and death.

d. Changes in heart rate or rhythm. Circulatory disturbances may also be associated with changes in heart rate or rhythm. Increased heart rate which is not associated with increased surgical stimulation can be caused by blood loss. Severe slowing of the heart can be caused by vagal stimulation, for example when traction is applied to the viscera, when ocular surgery is carried out, or when the nerve is handled during surgical procedures in the neck. This can be sufficiently severe to cause marked hypotension and can even result in cardiac arrest.

2. *Corrective action*

When attempting to correct signs of cardiovascular failure, it is helpful if some indication of the likely cause is obtained. However, whatever the causative factor, the following measures should be commenced:

- An immediate priority must be to ensure an unobstructed airway, preferably by endotracheal intubation. If intubation is possible, the animal should be ventilated with 100% oxygen, or at least this should be administered via a face-mask. If assisted ventilation cannot be provided using an anaesthetic circuit, intermittent compression of the chest wall should be commenced. In large animals, the air movements produced by this technique can be readily appreciated, but even in small rodents, gentle and rapid compression of the thorax between thumb and forefinger can result in effective ventilation.
- If complete cardiac arrest has occurred, external cardiac massage should be undertaken. In larger species, this is best achieved by placing the animal on its side and firmly compressing the chest over the region of the heart (just behind the point of the elbow). The compression should be applied smoothly and maintained for about half a second and at a rate of 60–70 compressions per minute. With smaller animals, the chest should be held between thumb and forefinger and the area over the heart compressed regularly and rapidly, about 90 times per minute. Even in rabbits and guinea pigs, some circulatory support can be maintained while other corrective measures are being instigated. Combining assisted ventilation and external cardiac massage in small rodents

requires practice and it is usually easier to compress all areas of the thorax simultaneously.

- After adequate ventilation has been established and cardiac massage attempted if it is necessary, an intravenous line should be inserted for drug and fluid therapy. To avoid the need to carry out emergency venepuncture, it is the author's practice to tape a suitable cannula in a superficial vein, either during or shortly after induction of anaesthesia, in all except the smallest animals. If anaesthetic overdose is suspected, either a specific antagonist or an analeptic such as doxapram should be administered. The use of drugs to restore stable cardiac rhythm and output requires considerable care and presupposes that arterial pressure and the ECG are being monitored. However, as an emergency measure, adrenaline (6 ml per 20 kg of 1:10000) should be given if asystole is suspected, or lignocaine (2 mg kg^{-1}) if the heart is fibrillating.

Arrhythmias will often respond to lignocaine, or to other antiarrhythmic agents such as bretylium (5–10 mg kg^{-1}). Complete heart block or low cardiac output can be treated by injection of atropine (0.02 mg kg^{-1}) and if required by the infusion of isoprenaline (5–20 μg kg^{-1} min^{-1}). If cardiac arrest has occurred, these drugs should be administered by intracardiac injection. After treatment of cardiac failure, sodium bicarbonate may be administered to correct the acidosis which is usually present. Although elaborate formulae are available for calculation of the dose required — see, for example, Lumb and Wynn Jones (1973) — a useful guide for emergency use is 1 mmol kg^{-1} of body weight. If cardiovascular failure has arisen primarily from hypovolaemia, maintenance of adequate ventilation and effective fluid therapy will rapidly restore a normal acid–base balance without the use of sodium bicarbonate.

The use of drugs to treat cardiac failure poses considerable problems to the inexperienced anaesthetist. Detailed descriptions of the techniques available have been given by Hensley and Martin (1990). All of the more sophisticated means of correcting and treating cardiac failure (including measures such as defibrillation) which are used in humans can be used in animals, and if high-risk procedures such as cardiac surgery are planned, then expert advice should be sought. A summary of emergency measures is given in Table 4.2.

C. Fluid balance

It is of vital importance to support the circulation by correcting any fluid imbalances, and hypovolaemia should always be considered a possible

TABLE 4.2
Basic guide for coping with cardiovascular emergencies, and infusion rates of some drugs commonly used for cardiovascular support (for detailed information, see Hensley and Martin, 1990)

For all cardiovascular problems:

- Administer 100% oxygen and ventilate, and turn off anaesthetic vaporizer or anaesthetic infusion
- If blood loss, transfuse (in order of preference)

  i. Whole blood
  ii. Haemaccel or Hespan (or equivalent products)
  iii. Lactated Ringer's solution

  If rapid blood loss has occurred replace estimated blood loss as quickly as possible, otherwise 10–15 ml kg^{-1} h^{-1}

- If low arterial pressure, administer:

  i. Dopamine infused at 5–10 μg kg^{-1} min^{-1}, then 1–5 μg kg^{-1} min^{-1} after volume replacement

  or

  ii. Adrenaline 1 μg kg^{-1} (0.2 ml per 20 kg of 1:10000), then infuse 0.05–0.5 μg kg^{-1} min^{-1}

- Cardiac arrest: start external cardiac massage

  a. Fibrillating

  i. Lignocaine 2 mg kg^{-1} i/v (4 ml of 1% per 20 kg), if no response administer 5 mg kg^{-1} i/v (10 ml of 1% per 20 kg) plus use defibrillation

  and/or

  ii. Bretylium 5–10 mg kg^{-1} plus use defibrillation
  iii. Adrenaline 30 μg kg^{-1} (6 ml per 20 kg of 1:10000)
  iv. Sodium bicarbonate 1 mmol kg^{-1} initially, reassess after blood gas analysis

  b. Asystole

  i. Adrenaline as above, repeat in 2 min if no response

  c. Heart block, bradycardia

  i. Atropine 0.02 mg kg^{-1} (0.6 ml per 20 kg)
  ii. If continued treatment needed: isoprenaline 5–20 μg kg^{-1} min^{-1}

Drugs for cardiac support:

Bretylium	5–10 mg kg^{-1}	To prevent dysrhythmias
Dobutamine	2.5–10 μg kg^{-1} min^{-1}	To increase cardiac output
Dopamine	1–5 μg kg^{-1} min^{-1}	To increase renal and mesenteric blood flow
	5–20 μg kg^{-1} min^{-1}	To increase heart rate and cardiac output, decrease renal blood flow and increase peripheral vascular resistance at higher doses
Lignocaine	30–70 μg kg^{-1} min^{-1}	To prevent dysrhythmias

primary cause of cardiovascular failure. Blood loss during surgery can be very gradual and assessment of the volume lost is frequently highly inaccurate. Whilst weighing all the swabs used will provide a rough estimate of blood loss, additional blood will have been lost by seepage into surgical wounds, body cavities and the surgical drapes. Additional losses of plasma occur by exudation both into traumatized tissues and also into the peritoneal cavity during prolonged abdominal surgery: approximately 100–200 ml per hour in humans (Wiklund and Thoren, 1985). A further depletion of the extracellular fluid (ECF) occurs through water loss by evaporation from the respiratory tract and from any surgical wounds and exposed viscera. As a routine, fluid should be replaced at a rate of 10 ml kg^{-1} of body weight per hour using either Hartmann's solution or 0.9% saline.

A healthy, unanaesthetized animal can withstand the rapid loss of 10% of its circulating volume. Once the loss exceeds 15–20% of circulating volume, signs of hypovolaemia and shock may develop. In an anaesthetized animal, many of the physiological mechanisms that act to maintain cardiovascular stability are depressed and hence less severe losses can still have serious effects. If blood loss exceeds 20–25% of the circulating volume, replacement with whole blood may be necessary. Smaller losses can be replaced by the infusion of crystalloid solutions or plasma volume expanders.

Blood can be obtained from a donor animal of the same species, and collected in acid citrate dextrose (ACD) solution (1 part ACD to 3.5 parts blood, using ACD from a human blood collection pack). It is preferable to use the blood within 4–6 hours, as platelet function and red cell viability is likely to be well maintained for this period. More prolonged storage at 4°C is possible, but the storage characteristics of blood from many animal species have not been properly evaluated. Although cross-matching will rarely be possible when dealing with laboratory animals, in the author's experience the incidence of adverse reaction to an initial transfusion appears low. Selection of donors of the same breed or strain as the recipient may help reduce the likelihood of transfusion reactions. Use of blood from a single individual, rather than pooled from a number of donors, will also help to reduce the risk of an adverse reaction. When using an inbred strain of rodent, there are obviously no problems of this nature.

Blood should be replaced at a rate of 10% of the calculated blood volume every 30–60 minutes. If severe and rapid haemorrhage has occurred, the estimated volume of blood lost should be transfused as rapidly as possible. If whole blood is unavailable, either previously stored plasma or a plasma volume expander such as Haemaccel (Hoechst) or Hespan (Du Pont) should be administered. It is important that stored plasma should be warmed to body temperature before infusion. If these fluids are unavailable, or if blood loss has been less severe, then Hartmann's solution or 0.9% saline should be administered, at the rate described for whole blood and at a volume of three to five times the estimated blood loss. Considerably greater volumes are needed because these crystalloids are distributed throughout the extracellular fluid, unlike blood, plasma and plasma volume expanders which remain in the circulatory system. Some controversy exists concerning the merits of crystalloids and plasma volume expanders for restoring the circulating volume following severe haemorrhage. Such controversy should not be a deterrent to the use of fluid therapy and it should be remembered that it is almost always better to give than to withhold fluid.

In small animals in which intravenous therapy is difficult, warmed 0.9% sodium chloride or Hartmann's solution can be administered intraperitoneally to correct intra-operative fluid loss. It is often particularly convenient to replace intra-operative water losses and anticipated post-operative deficits by the administration of 0.18% sodium chloride with 4% dextrose by subcutaneous injection at a rate of 10–15 ml kg^{-1}. These routes of administration result in slow absorption and are of no immediate value in treating cardiovascular failure.

D. Hypothermia

Hypothermia is a frequent cause of anaesthetic deaths. Hypothermia also prolongs recovery time from anaesthesia (Figure 4.3) and increases the potency of volatile anaesthetics (Regan and Eger, 1967). It is particularly important in small rodents and birds, but also occurs in larger species, especially during prolonged anaesthesia. Small mammals and birds lose heat rapidly because of their high ratio of surface area to body weight. The homeostatic mechanisms that control body temperature are depressed during anaesthesia and severe hypothermia can result. Hypothermia can develop rapidly in small animals; the author has recorded reductions in body temperature of 10°C in as little as 15–20 minutes in anaesthetized mice.

The fall in body temperature can be exacerbated by the flow over the animal of cold, dry gases from an anaesthetic machine. In addition, shaving the animal before surgery will remove insulating hair, and the use of cold skin disinfectants will cause further heat loss. During surgical procedures, exposure of the viscera and use of swabs soaked in cold saline, or the administration of cold intravenous fluids, will all cool the animal.

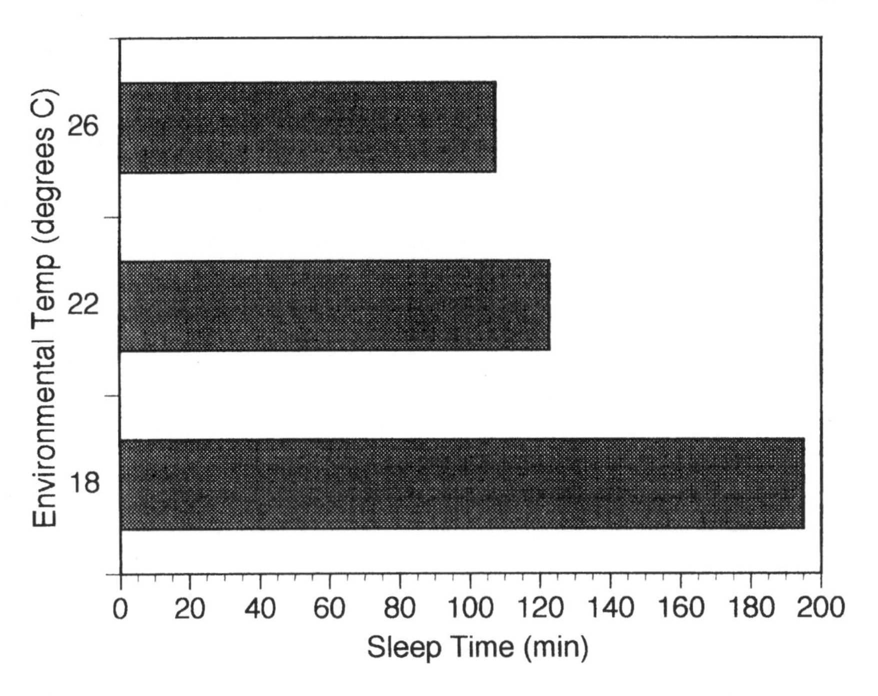

Fig. 4.3. Sleep time (duration of loss of righting reflex) in mice anaesthetized with pentobarbitone (60 mg kg^{-1}, i/p) and maintained at three different room temperatures. No attempt was made to prevent a fall in body temperature. Data from Lovell (1986c).

1. Preventive measures

Clearly, careful pre-operative and intra-operative management can reduce any fall in body temperature. Most animals will require some additional heating and should be insulated to minimize heat loss. Effective insulation can be provided either by wrapping the animal in cotton wool, followed by an outer wrapping of aluminium foil, or by using the bubble-packing which frequently forms part of the packaging of laboratory equipment, or other insulating materials. After the animal is wrapped in an insulating layer of material, a window can be cut to expose the operative field (Figure 4.4). When insulating small rodents, ensure that the tail is included in the wrapping, since heat loss from this part of the body can be considerable.

Additional heating can be provided by heat lamps and heating blankets, but care must be taken not to burn the animal. A thermometer placed next to the animal, or between the animal and a heating pad, will show whether excessive heat is being applied. The probe temperature should not exceed 40°C. It is possible to cause hyperthermia by overenthusiastic

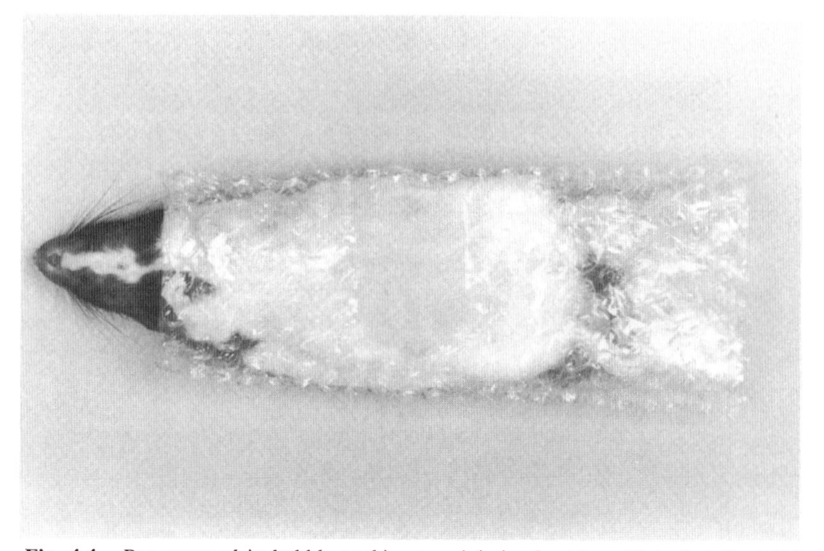

Fig. 4.4. Rat wrapped in bubble-packing to minimize heat loss. Note that the tail has been included in the wrapped area. A window has been cut to allow access to the abdomen for laparotomy.

or uncontrolled heating, and this can result in the death of the animal. To avoid such problems, and provide effective, well-controlled warming, it is preferable to use a thermostatically controlled heating blanket, regulated by the animal's body temperature using a rectal probe (Harvard Apparatus Ltd; International Market Supply; Appendix 7). If such a unit is not available, a simple heating pad or lamp can be used and, provided the animal's rectal temperature is monitored, can be switched on and off manually as required (Figure 4.5). It is important to switch on heating pads and lamps before they are required, to allow their temperature to stabilize and to prevent a period of inadequate heating when the pad or lamp is warming up. Thermostatically controlled pads can be set up with the probe in contact with the blanket, so that they reach body temperature before the animal is anaesthetized.

E. Vomiting and regurgitation

Vomiting or regurgitation of stomach contents may occur either during induction of anaesthesia or during the recovery period. It is a potentially serious problem and requires prompt treatment. Inhalation of gastric contents can produce immediate respiratory obstruction, asphyxiation and death, or lead to the development of aspiration pneumonia.

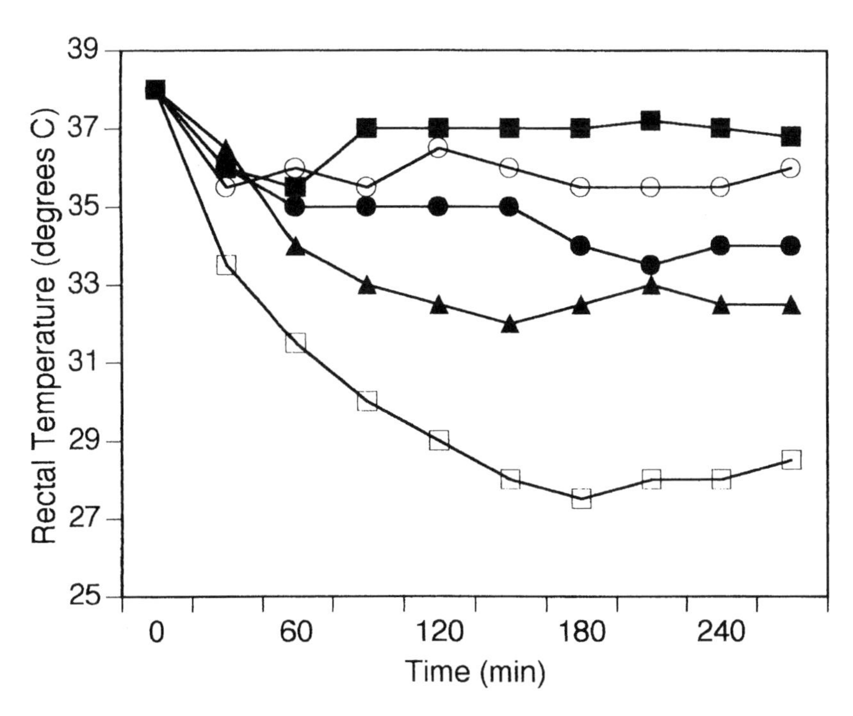

Fig. 4.5. The rate of fall of body temperature in five rats anaesthetized with fentanyl/fluanisone/midazolam, using a range of different techniques to reduce heat loss: thermostatically controlled pad (solid squares), non-thermostat controlled heating pad (open circles), 'Flectabed' (solid circles), bubble wrap (triangles), no insulation (open squares).

If vomiting is seen, the animal should immediately be placed in a head-down position and the vomit aspirated from the mouth and pharynx. If an effective suction apparatus is not available, one can be improvised from a large-diameter catheter and a 50 ml syringe. Since speed of reaction is of paramount importance, such apparatus should be available as standard equipment in the anaesthetic preparation room and recovery area.

If aspiration of vomit has occurred, oxygen should be administered and ventilation supported if respiratory distress develops. A broad-spectrum antibiotic should be administered, and corticosteroids given immediately by the intravenous route (30 mg kg^{-1} methylprednisolone). If administration is delayed, steroids may be of little benefit.

It is obviously preferable to reduce the incidence of vomiting and its associated problems by withholding food pre-operatively when appropriate (see Chapter 1) and by rapid endotracheal intubation of all animals whenever this is practicable.

5

Special techniques

I. CONTROLLED VENTILATION

Many anaesthetic agents depress respiration and this can lead to hypercapnia, hypoxia and acidosis. To maintain blood carbon dioxide and oxygen concentrations at more appropriate levels, it is often necessary to assist ventilation. If the thoracic cavity is opened, the normal mechanisms of lung inflation are disrupted and it is essential to ventilate the animal. It is not necessary to use a mechanical ventilator if a suitable anaesthetic circuit is in use (see Chapter 3), but use of a ventilator will often be more convenient than manually assisting ventilation. A mechanical ventilator will often allow the precise control of the duration of inspiration and expiration, of the volume of gas delivered to the lungs and of the pressure reached in the airway during inspiration. It is not necessary to administer a neuromuscular blocking agent (a muscle-relaxant or curare-like drug) in order to carry out artificial ventilation, but unless the animal is deeply anaesthetized or is hyperventilated to produce hypocapnia, spontaneous respiratory movements may occur and these may interfere with ventilation.

A. Neuromuscular blocking agents

Neuromuscular blocking drugs produce paralysis of the skeletal muscles. They may be used either to aid stable mechanical ventilation by blocking spontaneous respiratory movements or, more frequently, to provide more suitable conditions for surgery. If skeletal muscle tone is eliminated by using a neuromuscular blocking agent, exposure of a surgical site can be achieved more easily and with less trauma to the surrounding tissues. The neuromuscular blocking drugs in common clinical use are classified as either depolarizing or non-depolarizing agents.

Depolarizing agents, such as suxamethonium and decamethonium, act similarly to the normal transmitter at the neuromuscular junction, acetylcholine. They bind to muscle receptors and trigger a muscle contraction, but then produce a persistent depolarization, so preventing further muscle

TABLE 5.1
Dose rates for neuromuscular blocking agents (mg kg^{-1}, by intravenous injection)

Muscle relaxant	Mouse	Rat	Guinea pig	Rabbit	Cat	Dog	Sheep	Goat	Pig
Alcuronium	–	–	–	–	0·1	0·1	–	–	0·25
Atracurium	–	–	–	–	0·2	0·5	–	–	–
Gallamine	–	1	0·1–0·2	1	1	1	1	4	2
Pancuronium	–	2	0·06	0·1	0·06	0·06	0·06	0·06	0·06
Suxamethonium	–	–	–	0·5	0·2	0·4	0·02	–	2
Tubocurarine	1	0·4	0·1–0·2	0·4	0·4	0·4	0·4	0·3	–
Vecuronium	–	–	–	–	0·1	0·1	0·05	0·15	0·15

contractions. When drugs that act in this way are administered to an animal, generalized, disorganized muscle twitches are produced before complete skeletal muscle paralysis occurs.

Non-depolarizing or competitive blocking agents do not cause a muscle contraction before producing paralysis. Drugs in this group include tubocurarine, gallamine, pancuronium, alcuronium and vecuronium (Table 5.1). Because these agents act by competing with acetylcholine for receptor sites at the neuromuscular junction, their action can be reversed by increasing the local concentration of acetylcholine. This can be achieved by administering drugs such as neostigmine which block the activity of the enzymes that normally break down acetylcholine.

Neuromuscular blocking agents must be used with great care, since their administration prevents all movements in response to pain. It would be possible, but obviously inhumane, to carry out a surgical procedure on an animal which had been paralysed but was still fully conscious. It is for this reason that the use of neuromuscular blocking drugs in experimental animals is subject to careful control in many countries: for example, special permission is required to use these agents in the UK, and Institutional Animal Care and Use Committee review is required in the USA. Neuromuscular blocking agents are nevertheless extremely useful adjuncts to anaesthesia and enable, for example, the use of balanced anaesthetic regimens of an opiate, a hypnotic and a muscle relaxant to provide stable surgical anaesthesia. Dose rates of a number of different neuromuscular blocking agents are given in Table 5.1.

If neuromuscular blocking drugs are used, other methods of assessing the depth of anaesthesia must be adopted. As a preliminary step, the proposed anaesthetic regimen, excluding the neuromuscular blocking drug, should be administered to an animal of the same species and the proposed surgical procedure carried out. This will establish that the degree of analgesia and unconsciousness will be sufficient to allow surgery to be carried out humanely. Since considerable individual variation in response to anaesthesia occurs and some inadvertent alteration in the regimen can arise, for example due to equipment malfunction, it is also necessary to provide an independent assessment of the depth of anaesthesia. Several indicators of anaesthetic depth are of use. Despite muscle paralysis, twitching of muscles may occur in response to a major surgical stimulus and this indicates that the depth of anaesthesia is inadequate. In humans, pupillary size may alter in response to surgical stimulation, but this sign is of little value in most animals, particularly if atropine has been included in the pre-anaesthetic medication. Changes in blood pressure and heart rate are the most widely used indicators of adequacy of the depth of anaesthesia. Dramatic changes in heart rate or blood pressure are believed to indicate a depth of anaesthesia insufficient for the surgical procedures which are being undertaken. Despite their widespread acceptance, however, these parameters may not be completely reliable indicators of adequate anaesthesia (Whelan and Flecknell, 1992). Although most animals respond to surgical stimuli with a rise in blood pressure if inadequately anaesthetized, some animals may show a fall in pressure. If practicable, it is reassuring to allow the action of the muscle relaxant to subside periodically, so that the animal is again capable of responding to painful stimuli with voluntary movements. The degree of neuromuscular blockade can be monitored using a peripheral nerve stimulator. In any event, it is almost always feasible to delay administration of the relaxant until after the start of the surgical procedure, so that an initial assessment of the adequacy of the depth of anaesthesia can be obtained.

B. Mechanical ventilators

The majority of ventilators currently available have been designed primarily for use in humans, but many can be successfully adapted for animal use (Table 5.2). Ventilators are designed to achieve controlled ventilation of the animal's lungs by the application of intermittent positive pressure to the airway. This may be achieved either by delivering gas directly to the anaesthetic breathing circuit or, indirectly, by compressing a rebreathing bag or bellows, which in turn delivers gas to the animal.

A variety of techniques have been devised to control the delivery of gas to the patient and to determine the patterns of gas flow and gas pressure which occur during ventilation. It might be thought that all that was

TABLE 5.2
Characteristics of a selection of commercially available mechanical ventilators

	Tidal volume (ml)	Frequency (cycles per minute)	Minute volume (litres)	Type	Manufacturer
Oxford	40–350 (paediatric bellows) 150–1200	10–40	1.5–20	Volume-cycled, volume pre-set, constant flow generator	Penlon Ltd
Nuffield					
Series 400	50–2000	10–85	1–30	Time-cycled, volume and flow rate pre-set	Penlon Ltd
Series 200	50–2000	10–85	1–30	Time-cycled, volume and flow rate pre-set	Penlon Ltd
+ Newton valve	10–1300	10–85		Pressure-cycled	Penlon Ltd
Manley 4	200–1000	2–60	2–20	Time-cycled, pressure or volume pre-set	Blease Medical
Pulmovent MPP	1.5–12	10–60	2–20	Volume-cycled, flow pre-set	Ohmeda
Rodent Ventilator	0.2–30	18–150	0.004–4.3	Time-cycled, volume pre-set	Harvard Instruments
Rodent Ventilator	?2–250	10–200		Time-cycled, pressure pre-set	Harvard Instruments
Zoovent	?1–1500	3–600		Time-cycled, pressure pre-set	Hillmoor Electronic Consultants

required of a ventilator was to deliver the required volume of gas to the lungs at a predetermined rate. However, since the characteristics of the patient's lungs, changes in airway resistance and leaks in the anaesthetic circuit can all influence the volume of gas delivered, different techniques for terminating inspiration have been devised.

There are basically only two ways in which gas can be delivered during inspiration. The ventilator may deliver gas at a set pressure pattern: the pressure is determined by the machine, but the patient's airway characteristics will influence the volume of gas which is delivered, because the pressure reached in the airway depends upon the resistance to flow provided by the patient's lungs. If the ventilator is set to achieve a predetermined pressure, it will be reached earlier, and less gas will be delivered, if the patient's lungs provide a higher resistance to flow.

In contrast, a ventilator may be set to produce a fixed flow pattern, which will not be influenced by the patient's lung characteristics. Under these circumstances, the flow of gas will be constant but the pressure that develops in the airway will vary depending upon the patient's lung characteristics.

It is important how these two types of ventilators, termed 'pressure generators' and 'flow generators' respectively, are switched or cycled from inspiration to expiration. This can be achieved in several different ways, but the most frequently used method in animal ventilators is time cycling. Here, the change to expiration occurs after a pre-set time and is uninfluenced by changes in the patient's lungs. If a time-cycled ventilator is used, the pressure developed in the lungs, the gas flow and the volume delivered can all vary. The actual values of these variables will depend both upon the characteristics of the patient and upon whether the ventilator is a pressure or flow generator. If the power of the ventilator is very great relative to the resistance of the patient's lungs then, although time-cycled, the ventilator may in fact deliver a pre-set volume during inspiration.

An alternative to time-cycling is to determine the volume of gas that should be delivered during inspiration based on the animal's estimated tidal volume and change from inspiration to expiration when this volume has been delivered. In contrast, the changeover may be triggered not when a fixed volume of gas has been delivered, but when a predetermined airway pressure has been reached.

Once the lungs have been inflated and expiration begins, some mechanism must be used to trigger inspiration. In practice, only two techniques are used: the changeover can occur after a fixed time, or after airway pressure falls to a pre-set level.

The apparent complications introduced by the mechanics of ventilator design do have real effects on the patient. For example, if a fixed tidal volume is delivered, an increase in the animal's airway resistance will result in increased airway pressure; this may become excessive and cause damage to the lungs. In addition, there will be no compensation for leaks in the anaesthetic circuit so, if any leaks are present, there will be a fall in the volume of gas actually delivered to the lungs. A ventilator set to deliver gas until a pre-set pressure is achieved will compensate for leaks in the anaesthetic circuit, but an increase in the animal's airway resistance will result in a fall in the tidal volume delivered to the lungs. Ventilators that deliver gas at a fixed flow rate with a high generating pressure and that are either time- or volume-cycled, are unaffected by changes in the patient's lungs, but they may produce excessive airway pressures.

In selecting a ventilator for use in laboratory animals, the most important factor to be considered is the ability to ventilate a wide range of animal species. The range of tidal volumes which can be achieved is listed in Table 5.2, although it must be remembered that many of these rely on the presence of a leak-proof anaesthetic circuit and minimal compliance of circuit components such as connecting tubing. Additional features that may be needed are the ability to apply positive end-expiratory pressure (PEEP) and a facility for humidification of gases. It is also important to select a machine that is simple to use and is reliable and easy to maintain. The author's personal preference for a suitable multipurpose ventilator is the Nuffield 200 (Penlon Ltd, Appendix 7) which, when used with a Newton valve, can deliver tidal volumes ranging from 10 ml to 2000 ml. Alternatively, the Zoovent (Hillmoore Electronic Consultants, Appendix 7) can be used to ventilate a wide range of animal species. Both ventilators require a source of compressed gas to provide the driving power for the ventilator. If a piped gas supply is available, this does not present a problem. If small gas cylinders are used to drive the ventilator, then large numbers may be required during a prolonged anaesthesia. Since the driving gas does not reach the animal's lungs, a compressor delivering medical air is one possibility. Alternatively, a large cylinder of compressed air can be provided as the driving gas. If none of these solutions is thought practicable, then the mechanically driven ventilators manufactured by Harvard Apparatus can be used. For small rodents, either the Harvard volume pre-set rodent ventilator or the pressure-cycled ventilator may be used. The latter ventilator can also be used for larger species. These two ventilators are designed to ventilate the animal either with room air or with gas provided from an anaesthetic machine. If gas is supplied from an anaesthetic machine, then it is important that a pressure relief valve is incorporated into the circuit, between the fresh gas inflow and the ventilator, to prevent over-inflation of the animal's lungs. Most ventilators designed for clinical use incorporate this highly desirable feature.

TABLE 5.3
Suggested ventilation rates for laboratory animals. Tidal volumes of 10–15 ml kg^{-1} are normally required. Whenever possible the adequacy of ventilation should be assessed by monitoring the end-tidal carbon dioxide concentration or by arterial blood gas analysis

Species	Breaths per minute
Pig, dog (< 20 kg)	15–25
Pig (> 20 kg), sheep (> 20 kg), dog (> 20 kg)	10–15
Primates (> 5 kg)	20–30
Marmosets	40–50
Cat, rabbit (1–5 kg)	25–50
Guinea pig	50–80
Rat	60–100
Other small rodents	80–100

1. Practical considerations

To establish intermittent positive pressure ventilation (IPPV), calculate the required tidal volume (15 ml kg^{-1} body weight) and select a suitable respiratory rate. Generally, a rate slightly lower than the normal resting rate when conscious is adequate. Suggested initial ventilation rates are given in Table 5.3.

During IPPV, the heart and large veins in the thorax are compressed during inspiration, in contrast to the negative pressure which develops in the thorax during inspiration with spontaneous ventilation. The positive pressure produced during IPPV can reduce cardiac performance and cause a fall in blood pressure. To reduce this effect, inspiration should be completed in as short a period as possible, but must not be too rapid as this could result in high airway pressure. Conventionally, inspiratory to expiratory ratios are set to be 1:2, but ratios of 1:3 and 1:4 will often cause less cardiac depression, while maintaining inflation pressures below 20 cmH_2O. To monitor inflation pressures, if the ventilator is not equipped with a pressure monitor, place a needle in the inspiratory side of the anaesthetic circuit and attach it to a pressure transducer. In addition to checking that excessive pressures do not develop, by setting appropriate limits on the pressure monitor, it can act as an alert should the animal become disconnected from the breathing circuit or the ventilator malfunction.

II. LONG-TERM ANAESTHESIA

When animals are anaesthetized for only a short period, their ability to withstand numerous disruptions to their normal physiology will often enable them to survive even very poor anaesthetic technique. As the period of anaesthesia is extended, the adverse effects caused by poor technique become increasingly important. Similarly, the undesirable side-effects of many anaesthetic drugs become more apparent and a considerably higher standard of intra-operative care becomes necessary. 'Long-term anaesthesia' is an arbitrary term, but here it is used to describe anaesthesia lasting longer than 60 minutes.

There is little practical difference between anaesthesia from which the animal will be allowed to recover and that in which the animal will be killed at the end of the procedure. Prolonged, non-recovery anaesthesia, often undertaken to enable the study of physiological mechanisms or drug metabolism, usually requires stable anaesthesia with minimal depression of the various body systems. However, since recovery is not required, cumulative effects of drugs become less important, provided that physiological stability can be maintained.

A. Injectable agents

1. Short-acting injectable anaesthetic agents

It may be thought that the simplest method of prolonging anaesthesia would be to give repeated doses of an injectable anaesthetic. Two problems arise if this approach is adopted. Giving intermittent doses of the drug will cause the depth of anaesthesia to vary considerably, although this can be overcome by administering it as a continuous infusion so that steady plasma concentrations of the anaesthetic are maintained. A second problem arises because of the pharmacodynamics of the anaesthetic agent. Following an initial injection of, for example, a barbiturate, the blood concentration of the drug rises rapidly and the concentration in tissues with high relative blood flows, such as the brain, also increases rapidly. Redistribution of the drug to other body tissues then follows, with equilibrium with body fat occurring most slowly. As this redistribution occurs, the concentration of drug in the brain falls. Recovery from the anaesthetic effects of the drug is primarily due to this redistribution, rather than to drug metabolism or excretion. If a second dose of anaesthetic is given, redistribution occurs more slowly, since the body tissues already contain some of the drug, and the duration of anaesthesia is prolonged. Repeated doses will have progressively greater effects. In addition to extending the duration of surgical anaesthesia, this approach also prolongs the sleeping time following anaesthesia. If the animal does eventually wake up, the residual effects of the drug may persist for 24–48 hours. For this reason, repeated incremental doses of drugs such as the barbiturates are not an ideal way of prolonging anaesthesia. The cumulative effects of different types of anaesthetic agent do vary considerably and some, such as alphaxalone/alphadolone, are rapidly metabolized following their administration. These drugs can be used to produce prolonged periods of anaesthesia without causing greatly extended recovery times.

Whichever injectable anaesthetic is used, it is preferable to administer incremental doses of the drug by the intravenous route, so that its effects on the depth of anaesthesia will be seen rapidly and be readily adjusted. Administration by other routes is possible with some drugs, but the depth of anaesthesia will vary less predictably.

a. Barbiturates. Recovery following repeated doses of barbiturates is prolonged, so the use of these drugs for procedures from which the animal is expected to recover consciousness is not recommended. Incremental doses of barbiturates can be used for non-recovery experiments, but the hypotension and respiratory depression that may result can cause serious problems. In addition, the depth of anaesthesia will also vary considerably and may be insufficient to allow surgical procedures to be undertaken in rodents and rabbits. In larger species, continuous infusion of pentobarbitone or repeated administration can be used successfully for long-term anaesthesia, but recovery is very prolonged and often associated with long periods of involuntary excitement and ataxia. The drug is therefore best reserved for use in non-recovery procedures.

b. Alphaxalone/alphadolone. Even after several hours of continuous infusion of alphaxalone/alphadolone (Saffan, Mallinkrodt Veterinary), recovery is rapid (Cookson and Mills, 1983) and it is useful for producing long-term anaesthesia in primates, cats, sheep, pigs, rats, mice and guinea pigs. In the rabbit the poor degree of analgesia and the respiratory depression that is produced at high dose rates limits the usefulness of this drug unless ventilation is assisted. Approximate infusion rates are given in Table 5.4. Initially, an induction dose of the drug should be given, followed by a continuous infusion at the rate quoted. The animal should be monitored to ensure the depth of anaesthesia is appropriate and the infusion rate increased or decreased as required. Stable anaesthesia is usually achieved within 30 minutes and will be established more rapidly as experience is gained with a particular species and strain of animal.

TABLE 5.4
Suggested regimens for total intravenous anaesthesia for long-term anaesthesia. Note that regimens using opioids often require intermittent positive pressure ventilation to maintain adequate ventilation

Species	Regimen
Cat	Alphaxalone/alphadolone 9–12 mg kg^{-1} i/v, then 0.2–0.3 mg kg^{-1} min^{-1} i/v Propofol 7.5 mg kg^{-1} i/v, then 0.2–0.5 mg kg^{-1} min^{-1}
Dog	Propofol, 5–7.5 mg kg^{-1} i/v, then 0.2–0.4 mg kg^{-1} min^{-1}; addition of alfentanil (2–3 μg kg^{-1} min^{-1}) enables propofol rate to be reduced to 0.14–0.18 mg kg^{-1} min^{-1} Midazolam 50–100 μg kg^{-1} i/v and alfentanil 10–20 μg kg^{-1}, then midazolam 5 μg kg^{-1} min^{-1} and alfentanil 4–5 μg kg^{-1} min^{-1}
Mouse	Alphaxalone/alphadolone 15–20 mg kg^{-1} i/v, then 0.25–0.75 mg kg^{-1} min^{-1} i/v Propofol 26 mg kg^{-1} i/v, then 2–2.5 mg kg^{-1} min^{-1}
Non-human primate	Alphaxalone/alphadolone 10–12 mg kg^{-1} i/v, then 0.3–0.6 mg kg^{-1} min^{-1} i/v Propofol 7–8 mg kg^{-1} i/v, then 0.2–0.5 mg kg^{-1} min^{-1} i/v
Pig	Alphaxalone/alphadolone 2 mg kg^{-1} i/v after ketamine (10 mg kg^{-1} i/m) pre-med, then 0.1–0.2 mg kg^{-1} min^{-1} i/v Propofol 2–2.5 mg kg^{-1} i/v after ketamine (10 mg kg^{-1} i/m), then 0.1–0.2 mg kg^{-1} min^{-1}, addition of alfentanil (20–30 μg kg^{-1} i/v), then 2–5 μg kg^{-1} min^{-1}
Rabbit	Fentanyl/fluanisone 0.3 ml kg^{-1} i/m and midazolam (1–2 mg kg^{-1} i/v), then fentanyl 2–5 μg kg^{-1} min^{-1}
Rat	Alphaxalone/alphadolone 10–12 mg kg^{-1} i/v, then 0.2–0.7 mg kg^{-1} min^{-1} i/v Propofol 10 mg kg^{-1} i/v, then 0.5–1.0 mg kg^{-1} min^{-1}
Sheep	Alphaxalone/alphadolone 2–3 mg kg^{-1} i/v, then 0.1–0.2 mg kg^{-1} min^{-1} i/v

c. **Propofol.** Propofol has been shown to have little cumulative effect, and this drug has been used to provide prolonged anaesthesia in a number of species (Hall and Chambers, 1987; Blake *et al.*, 1988; Flecknell *et al.*, 1990b; Brammer *et al.*, 1992; Robertson *et al.*, 1992; Aeschbacher and Webb, 1993b). Maintenance of full surgical anaesthesia with propofol alone requires high infusion rates, and this may result in more prolonged recovery times in some species. It is often preferable to supplement propofol anaesthesia with a potent opioid such as alfentanil. This allows lower infusion rates of propofol to be used, and since the opioid can be reversed using a partial agonist such as nalbuphine, recovery is generally rapid (Flecknell *et al.*, 1990b). When using higher doses of opioids with propofol, respiration may be depressed, and it is preferable to connect the animal to a ventilator. This does not represent a significant disadvantage since the physiological state of the animal during long-term anaesthesia is almost invariably improved by using intermittent positive pressure ventilation.

Infusion with propofol will result in marked lipaemia, because the commercial preparation of the drug is as an emulsion in soya bean oil. This appears to have little clinical significance, but potentially could interfere with some research protocols.

d. Ketamine. Ketamine is metabolized moderately rapidly and incremental doses can be given to extend the period of anaesthesia. The recovery rate is prolonged after repeated administration and severe respiratory depression can occur. If repeated doses or a continuous intravenous infusion are to be used, it is usually necessary to monitor respiration and to have facilities for mechanical ventilation readily available. It is a poor anaesthetic and analgesic in most rodents (see Chapters 3 and 7). It will almost always be necessary to administer the drug in combination with medetomidine or xylazine or a sedative such as diazepam to eliminate the muscle rigidity which is produced when using ketamine alone (see Chapter 3).

e. Etomidate and metomidate. Etomidate and metomidate are potentially useful drugs for the production of long-term anaesthesia. Rapid metabolism of the drugs occurs and continuous infusion results in only a mild cumulative effect on recovery time. The drugs suppress adrenal cortical activity when used in this way (Fellows *et al.*, 1983; Wagner and White, 1984) and this side-effect must be considered if they are to be administered to experimental animals. Neither drug produces sufficient analgesia to allow surgical procedures to be carried out and an opiate such as fentanyl should be administered to produce full surgical anaesthesia. Metomidate and etomidate have relatively little depressant effect on the cardiovascular system (Nagel *et al.*, 1979; Kissin *et al.*, 1983), and when used with an opiate they produce good surgical anaesthesia and maintain stable cardiac function.

f. Neuroleptanalgesic combinations. These combinations, when administered with a benzodiazepine, produce excellent surgical anaesthesia in rodents and rabbits (Chapters 3 and 7), but several problems arise when they are used to produce long-term anaesthesia.

Neuroleptanalgesic preparations consist of a mixture of an opiate analgesic and a potent tranquillizer. If repeated doses of a fixed-dose preparation are given, relative overdosage with the tranquillizer may occur since it usually has a longer duration of action than the analgesic component. In practice, this seems to have only a minor effect in prolonging recovery times when fentanyl/fluanisone (Hypnorm, Janssen) is administered to rats, mice and rabbits. The problem can be avoided by inducing anaesthesia with the neuroleptanalgesic combination and a benzodiazepine and then maintaining anaesthesia with an infusion of an analgesic alone (e.g. fentanyl). Repeated doses of the benzodiazepine are required only every 4–6 hours.

A more serious side-effect can be the respiratory depression that is frequently seen following the administration of neuroleptanalgesics. This is not usually so severe as to require assisted ventilation, but a moderate hypercapnia and acidosis will develop. During prolonged anaesthesia it is preferable to assist ventilation by using a mechanical ventilator.

The neuroleptanalgesic combination that has been used most extensively in rodents and rabbits is fentanyl/fluanisone (Hypnorm), together with diazepam or midazolam. The combination of fentanyl and dropiderol (Innovar Vet) differs in its effects, and is generally unsuitable when used in this way. Repeated doses of Hypnorm can be given every 20–30 minutes to maintain anaesthesia in rats, mice, rabbits and guinea pigs, following an initial dose of Hypnorm together with a benzodiazepine, although the depth of anaesthesia will fluctuate markedly. It is preferable to administer a 1:10 dilution of the drug by intravenous infusion. In other species, such as the dog, a combination of fentanyl or alfentanil and midazolam can be used to provide long-lasting surgical anaesthesia (Flecknell *et al.*, 1989b), (Table 5.4).

g. Delivery techniques. It is preferable to use an infusion pump to deliver the drug, although a diluted solution can be administered using a paediatric burette. These burettes deliver 60 drops per millilitre so allowing greater control over the infusion rate than the standard adult-type apparatus. It is important to remember that this method of delivery will be very sensitive to partial occlusion of the catheter caused, for example, by changing the position of the animal or by the formation of thrombi in the catheter tip. In addition, gradual movements of the drip control device will result in changes in the infusion rate. For these reasons, if simple drip sets are used for infusion of anaesthetics they must be monitored carefully and frequently. The degree of control over the infusion rate can be improved by use of a drip-rate controller which varies the infusion rate by changing the diameter of the drip tubing. The simplest and least expensive of these are disposable devices, but greater accuracy can be achieved by using electronically controlled devices which measure the flow rate by counting the drip rate of the fluid. Electronic controllers also incorporate an alarm to alert the anaesthetist to cessation of flow, caused (for example) by occlusion of the catheter or exhaustion of the fluid reservoir. All of these devices will fail should occlusion of the catheter occur.

Catheter occlusion is less likely to occur when using an infusion pump because the driving pressure generated by the pump tends to maintain catheter patency. Infusion pumps are available from a number of different manufacturers and vary considerably in their suitability for laboratory animal anaesthesia. When selecting a pump, the most important considerations are as follows:

1. Can the pump deliver a wide range of different infusion rates? Ideally, the range should extend from microlitres per hour to millilitres per minute, and be capable of being varied in small increments — pumps that can only double or halve the rate of infusion should be avoided.
2. Can the pump accept syringes of different sizes, from a variety of different manufacturers?
3. Is the pump fitted with an occlusion alarm, a 'syringe empty' alarm and a power failure or low battery alarm?
4. Is a battery back-up provided in case of failure of the electrical supply or inadvertent disconnection?
5. Is the pump small, light, portable, easily cleaned and robust?
6. Is it easy to set and vary the infusion rate and are any necessary calculations carried out by the pump's own microprocessor?
7. Is an interface provided for microcomputer control of pump operation and is appropriate software provided to simplify this control process?

As with other anaesthetic apparatus, it is helpful to obtain potentially suitable infusion pumps on a trial basis, to ensure that they fulfil most of these requirements, and prove reliable during routine operation.

Whichever method of infusion is used, initial adjustments to the infusion rate will be required after induction of anaesthesia, but once experience has been gained with a particular anaesthetic regimen, stable infusion rates and depth of anaesthesia can be established relatively rapidly. Initially the infusion rate should be based on the pharmacodynamics of the anaesthetic drugs used, although it must be appreciated that even when full details of these have been published for a particular species, considerable between-animal variability will occur. Nevertheless, if the drug has been well characterized, then the required infusion rate can be estimated from its volume of distribution (the theoretical space in the body available to contain the drug) and its rate constants (Mather, 1983).

A simple analogy may be helpful to those unfamiliar with pharmacokinetics. If a particular concentration of dye is needed in a sink filled with water, the concentration of dye needed to be added is equal to the volume of water in the sink multiplied by the target concentration. If the sink's taps are turned on, and the plug is pulled out, then the situation becomes more complicated. In these circumstances, dye must be added continuously to maintain the desired concentration. If dye is added at a constant rate, then eventually a situation will be reached where the rate at which it is removed is equal to the rate at which it is added, a situation known as the 'steady state'. These same considerations apply to continuous intravenous infusion of anaesthetics. If an anaesthetic drug is infused at a constant rate, eventually the rate of removal from the plasma will equal the rate of infusion. Under these circumstances:

$$\text{Maintenance infusion rate} = \text{clearance} \times \text{plasma concentration}$$

Unfortunately this is an oversimplification for many anaesthetics, as their kinetics are better described by more complex models than a single sink, or single compartment. Many intravenous anaesthetic agents are characterized by one compartment with rapid distribution and elimination, and one or more compartments with slower equilibration and elimination times. The rapid distribution compartment is often thought to represent the blood and other well-perfused tissues. Drugs with these characteristics will have a single rapid half-life, and one or more slower half-lives. These are calculated from the rate of fall of the plasma concentration after administration of a single intravenous dose of the compound. The half-life of a compound is the time taken for its plasma concentration to fall by 50%.

If the anaesthetic is infused at a constant rate, it will require 4–5 half-lives to achieve a steady state. The more usual alternative is to administer an initial loading dose to induce anaesthesia, followed by a constant infusion. The problem with this latter technique is that if the drug's pharmacokinetics are best represented by a multicompartment system, then the plasma concentration will fall rapidly as redistribution to other compartments occurs. If additional anaesthetic is given rapidly to compensate, then dangerously high plasma levels can be attained. One method of estimating the loading dose required to achieve a steady plasma concentration rapidly has been described by Nowich (1977) as:

$$\text{Loading dose} = \text{maintenance infusion rate} \times \text{half-life}/0.693$$

The slow half-life is used with anaesthetics modelled using multicompartment systems. This method can result in high plasma concentrations, so if the required loading dose exceeds the recommended safe induction concentration, a safer approach is to multiply the maintenance rate by the half-life of the anaesthetic. This would prolong the time taken to reach a steady state, but would reduce the chance of overdose. An alternative approach is to use two infusions, an initial rapid rate followed by a slow maintenance rate, with the rates determined by the drug clearance and the half-life (Wagner, 1974).

If the half-life of the drug is not known, but practical experience has been gained by giving intermittent injections of the drug to maintain anaesthesia, then the total quantity of drug given in a specific time period can be used to estimate an infusion rate. It may also be possible to extrapolate half-lives of anaesthetics between species, as has been suggested for antibiotics, using the relationship of body weight $(kg)^{0.25}$ (Morris, 1995).

2. *Long-acting injectable anaesthetic agents*

An alternative to administering repeated doses or a continuous infusion of short-acting agents is to select an anaesthetic with a prolonged duration of action.

a. Urethane. Urethane has been widely used for the production of long periods of anaesthesia in a range of laboratory species and is reported to cause minimal depression of the cardiovascular and respiratory systems (Buelke-Sam *et al.*, 1978). It should be noted, however, that this cardiovascular stability is in part due to sustained sympathetic nervous system activity, associated with high circulating levels of adrenaline and noradrenaline (Carruba *et al.*, 1987).

Urethane has been reported to be both mutagenic and carcinogenic (Field and Lang, 1988) and, although the experimental studies on its potency are somewhat limited, it is now widely regarded as a potential hazard to staff. Before using urethane it is suggested that other anaesthetics are assessed and an alternative regimen used whenever possible. If it can be shown that the use of other anaesthetics would frustrate the purpose of the experiment and a decision is taken to use urethane, it is recommended that it be treated as a moderate carcinogen. In most institutes there will be guidelines for the safe handling of such materials and these will usually include the use of gloves and face-masks when handling the substance and use of fume cupboards or similar cabinets for preparing solutions from the dry powdered drug.

If such precautions are adopted, urethane can be used in a reasonably safe manner and is a valuable anaesthetic for providing long-lasting, stable surgical anaesthesia. In view of its carcinogenic action in rodents, it seems inadvisable to allow animals to recover from urethane anaesthesia.

b. Alpha-chloralose. When prolonged anaesthesia is required and surgical interference is to be kept to a minimum, use may be made of alpha-chloralose. This drug is a hypnotic and has little analgesic action, although this varies considerably between different species and strains of animal and in some individuals surgical anaesthesia is produced. It is usually necessary to administer a short-acting anaesthetic, such as methohexitone or alphaxalone/alphadolone, while carrying out any surgical procedures, following which alpha-chloralose can be administered to produce long-lasting, light anaesthesia. This drug is belived to be particularly valuable for studies of the cardiovascular system, since the various autonomic reflexes are well maintained (Holzgrefe *et al.*, 1987).

Alpha-chloralose is prepared by heating the powdered drug in water at 60°C to form a 1% solution. Care must be taken to avoid boiling the drug and the solution must be cooled to 40°C before administration. Administration in propylene glycol improves solubility and is claimed to reduce problems of acidosis associated with administration of alpha-chloralose (Shukla and Shukla, 1983). The onset of action following intravenous administration is about 15 minutes so, even if surgical procedures are not to be undertaken, it is preferable to administer a short-acting drug to induce anaesthesia, followed by the alpha-chlorolose. The drug produces 8–10 hours of light anaesthesia in most species (see Chapter 7 for dose rates).

In addition to the use of alpha-chloralose as the sole anaesthetic, various combinations with urethane have been described, which aim to reduce the quantity of urethane required, and provide improved analgesia compared with alpha-chloralose alone (Korner *et al.*, 1968; Sharp and Hammel, 1972; Hughes *et al.*, 1982). These combinations do not circumvent one of the main difficulties of using urethane — its classification as a carcinogen. An alternative approach is to provide additional analgesia using opioids in combination with alpha-chloralose (Rubal and Buchanan, 1986).

c. Inactin. Inactin, a thiobarbiturate, has been widely used as a long-acting anaesthetic in the rat. In this species it produces surgical anaesthesia in some strains of animal, with well-maintained systemic arterial pressure (Buelke-Sam *et al.*, 1978). Despite near-normal arterial pressure, blood flow to specific organs may be significantly reduced, owing to depressed cardiac output (Walker *et al.*, 1983), and it should not be assumed that cardiovascular function is normal. Nevertheless, this agent can be extremely useful for producing 3–4 hours of general anaesthesia.

B. Inhalational agents

There is very little practical difference between administering an inhalational agent for 30 minutes, or for 8 hours. Once the animal has been anaesthetized, it will remain anaesthetized if supplied with an appropriate maintenance concentration of anaesthetic. After 1–2 hours, the maintenance concentration of anaesthetic can usually be reduced, and further reductions may be possible, particularly if no further surgical stimulus is given. Great care must be taken, however, if neuromuscular blocking agents are used, since signs of lightening of anaesthesia may be less obvious.

The major problems in using volatile anaesthetics for prolonged periods are usually due to poor anaesthetic techniques or to equipment failure. It is essential to select an anaesthetic circuit with minimal dead space and minimal circuit resistance. It is also important to ensure that adequate supplies of compressed gas cylinders and anaesthetic agent are available (Appendix 2). It is not uncommon to exhaust four or five gas cylinders during a prolonged period of anaesthesia, particularly if the compressed gas source is also used to drive a mechanical ventilator.

C. Nitrous oxide/relaxant anaesthesia

An alternative regimen which has been claimed to produce prolonged, stable anaesthesia is the use of neuromuscular blocking agents and nitrous oxide. This type of regimen is totally unacceptable for use in animals, since it will not provide sufficient depth of anaesthesia to allow surgical procedures to be carried out humanely, or even produce loss of consciousness reliably. Confusion appears to have arisen because of the use of relaxant/nitrous oxide regimens in humans. The potency of inhalational anaesthetics is commonly expressed as the minimum alveolar concentration (MAC). This is the alveolar concentration of the anaesthetic required to prevent movement in response to a defined painful stimulus (e.g. clamping a haemostat on a digit or the tail) in 50% of the population. In humans, the MAC value of nitrous oxide is 95% and when used alone it can produce loss of consciousness and moderate analgesia. In animals, the MAC value ranges from 150% to 220% (Steffey *et al.*, 1974; Weiskopf and Bogetz, 1984; Tranquilli *et al.*, 1985; Mahmoudi *et al.*, 1989) (see Table 3.5). When used alone, it cannot produce sufficient analgesia for even the most superficial surgical procedure and does not even appear to produce loss of consciousness in most species. The occurrence of awareness during relaxant/nitrous oxide anaesthesia is a recognized problem in humans

(Breckenridge and Aitkenhead, 1983). In animals, the low potency of nitrous oxide will almost invariably result in awareness in animals paralysed by neuromuscular blocking agents. Even if surgical procedures are not carried out, this is likely to cause considerable distress to the animal.

There have been suggestions that following completion of surgery using conventional anaesthetic regimens together with a neuromuscular blocking agent, the surgical wounds can be infiltrated with local anaesthetic, and anaesthesia continued using nitrous oxide alone. Besides the problem of awareness discussed above, the local anaesthetic agents used (lignocaine) have a short duration of action and movements at the sites of surgical incision could result in pain once the drug's effects had disappeared. For these reasons the technique is considered unacceptable by the author.

D. Management of long-term anaesthesia

As mentioned earlier, all the potential problems and adverse effects of anaesthesia assume greater importance during a prolonged period of unconsciousness. All the factors discussed in Chapter 4 should be considered, but some additional care will also be required.

1. Respiratory and cardiovascular function

Because all anaesthetics depress respiration to some extent, it is advisable to administer oxygen to all animals that are anaesthetized for prolonged periods. It is also preferable to intubate the animal and provide facilities for mechanical ventilation. Mechanical ventilation is essential if an attempt is to be made to provide stable blood-gas concentrations during anaesthesia.

In a spontaneously breathing animal, the depression in respiration caused by anaesthesia produces a rise in $P\text{CO}_2$. The rebreathing which occurs because of equipment dead space will also contribute to this rise. In most animals, this will be of little clinical consequence for an hour or so; after this period problems associated with hypercapnia and acidosis begin to develop.

One approach to control this problem is periodically to assist ventilation by manual compression of a reservoir bag or by intermittent occlusion of a T-piece. Since, during anaesthesia, respiratory drive is influenced strongly by blood-oxygen tension, temporarily increasing oxygen tension by ventilating with 100% oxygen can produce a short period of apnoea. In addition, an elevation in blood carbon dioxide tension increases the release of catecholamines and these have a stimulating effect on the cardiovascular system. If carbon dioxide tensions are reduced by a short period of assisted ventilation, this can produce a fall in cardiac output. The previously elevated carbon dioxide tensions will also have produced a peripheral vasodilation and this will persist during the period of assisted ventilation. The fall in cardiac output, coupled with the persistent vasodilation, may produce a period of hypotension. These problems are best avoided by ensuring that the animal's initial ventilation is adequate and if the minute volume is low, mechanical ventilation should be started early in the course of the anaesthetic and maintained throughout its duration.

A second problem occurring during long-term anaesthesia is caused by a build-up of bronchial secretions, which can block small airways. Use of atropine or glycopyrrolate can help to reduce the quantity of these secretions, but partial airway obstruction may still occur. A freer flow of bronchial mucus can be produced by humidifying the inspired gas mixture. Although purpose-made nebulizers are to be preferred, simply bubbling gases through a temperature-controlled water bath appears satisfactory during controlled ventilation of rats. When anaesthetizing larger species, disposable humidifiers can be incorporated into the anaesthetic circuit.

During prolonged periods of anaesthesia, metabolic acidosis may gradually develop. This can only be detected by monitoring arterial blood gases and pH, and may be corrected by administration of sodium bicarbonate. Gradual fluid loss from surgical wounds and the respiratory tract, and by urine formation, should be replaced by continuous infusion of balanced electrolyte solutions (Hartmann's solution). During prolonged anaesthesia it is useful to monitor urine production by catheterizing the bladder, to verify continued renal function.

2. *Hypothermia and fluid balance*

In addition to the problems of providing adequate ventilation, all the monitoring and management techniques described in Chapter 4 assume much greater importance. Particular attention should be given to the prevention of hypothermia, and fluid balance should be carefully monitored. Although fluid deficit is the major concern during prolonged surgery and anaesthesia, it is important to avoid fluid overload, caused by a combination of enthusiastic intravenous therapy and the infusion of large volumes of anaesthetic drugs. As a general guide, total fluid infusions of up to 10% of circulating volume per hour (7 ml kg^{-1} h^{-1}) are well tolerated by most animals.

3. Posture

A problem which is of considerable importance in human anaesthetic practice, but which receives little attention in laboratory animal anaesthesia, is the damage that can be caused to muscles and nerves by the adoption of an abnormal position during anaesthesia. To avoid unnecessary post-operative discomfort, try to ensure that the animal is placed in as normal a posture as possible. Avoid tying out the limbs and instead use positioning pads. Try to protect pressure points such as the elbow, hock and the wings of the pelvic bones. If possible, change the animal's position and massage these pressure points every 1–2 hours. The eyes should be protected, preferably by taping them shut, to avoid corneal desiccation. The addition of a bland ophthalmic ointment (e.g. Visco-Tears, Ciba, or Lacri-lube, Allergan) is also of use and this has the advantage of providing some protection during post-operative recovery. The mouth, nose and pharynx can become blocked with viscous secretions and these should be removed using gentle suction before the animal is allowed to recover.

III. ANAESTHESIA OF PREGNANT ANIMALS

Pregnant animals require special care when anaesthetized. Consideration must be given both to the adverse effects of anaesthesia on the mother and to its effects on the fetus. The increasing size of the fetuses in the last third of pregnancy leads to an increase in abdominal pressure and consequent interference with respiratory movements. This may be of minimal importance under normal circumstances, but may be of considerable significance when the mother is placed in an abnormal posture during anaesthesia. The pressure of the uterine contents on the abdominal blood vessels may also interfere with venous return.

To minimize these problems, care should be taken to avoid maintaining the animal in dorsal recumbency. Wherever possible, position the animal so that it is lying on one side. It may be advisable to carry out endotracheal intubation and assist ventilation, particularly during the last third of pregnancy. Good anaesthetic practice, such as providing oxygen by face-mask to the animal before induction, assumes greater importance, but care must also be taken to reduce stress. Pregnant animals should not be starved before inducing anaesthesia, as this can have adverse metabolic effects both on mother and fetus. The fetus is extremely sensitive to changes in maternal acid–base balance caused, for example, by hypercapnia. Maternal hypotension can seriously reduce placental blood flow and cause the fetus to develop hypoxia. The fetus in many species is very susceptible to hypothermia, and special care must be taken both to maintain maternal body temperature, and to keep the fetus warm if it is exteriorized.

A. Placental transfer of anaesthetic drugs

Most commonly used anaesthetic drugs cross the placenta. This may be advantageous in providing some degree of anaesthesia in the fetus to allow surgical procedures to be undertaken. Conversely, the drug may have serious acute or long-term effects on the fetus. If the fetus is to be delivered by Caesarean section, residual effects of the anaesthetic drug may cause sedation, respiratory depression and cardiovascular system depression.

B. General recommendations

The choice of a particular anaesthetic regimen will depend upon the type of experimental procedures which are to be undertaken, but the following advice can be offered as a general strategy for anaesthetizing pregnant animals.

1. Use a balanced anaesthetic regimen to reduce adverse effects such as hypotension in the mother and so minimize hazards to the fetus.
2. Use local and regional anaesthesia if possible, but this must be balanced against welfare considerations (see Chapter 3).
3. Carry out surgical and other techniques as rapidly as possible to reduce the duration of anaesthesia.
4. Whatever the anaesthetic regimen, maintain good oxygenation of the mother and minimize the incidence of hypercapnia by assisting ventilation.
5. Maintain blood pressure with intravenous fluids and plasma volume expanders whenever necessary.
6. Monitor maternal blood glucose levels and correct any hypoglycaemia that may develop.
7. If the fetus is being delivered by Caesarean section, and narcotic analgesics have been administered to the mother, administer naloxone to the neonate to reverse any respiratory depression caused by these agents. Irrespective of the anaesthetic regimen used, administration of doxapram to stimulate respiration can be helpful.

C. Fetal surgery

If surgical procedures are to be undertaken on the fetus, care must be exercised to ensure that it is adequately anaesthetized. The stage of gestation at which the central nervous system is sufficiently well developed to respond to painful stimuli varies in different species, and expert advice as to the likely stage of development of the fetus should be sought before commencing work.

If the fetus is sufficiently developed to respond to painful stimuli, then pain must be prevented by means of an appropriate anaesthetic.

The simplest approach is to anaesthetize the mother to a sufficiently deep level so that the fetus is also anaesthetized. This is most readily achieved by using a volatile general anaesthetic such as methoxyflurane. Maintenance of this depth of anaesthesia may be undesirable, however, because of possible adverse effects, such as cardiovascular system depression, on mother and fetus. A single dose of an anaesthetic administered by the intravenous route, although sufficient to produce anaesthesia in the mother, is very unlikely to produce anaesthesia in the fetus.

The most widely used alternative to deep general anaesthesia is to infiltrate the surgical site on the fetus with local anaesthetic. If carried out carefully, this technique appears suitable for procedures such as cannulation of superficial blood vessels. It is unlikely to be sufficiently effective for more major surgery such as laparotomy or thoracotomy.

IV. ANAESTHESIA OF NEONATES

Neonatal animals have an increased susceptibility to hypothermia and may also have poor pulmonary and circulatory function. They frequently have low energy reserves, which can lead to problems during the recovery period. In addition, any period of hypoxia can lead to rapid depletion of hepatic glycogen stores and result in hypoglycaemia. Neonates have a reduced capacity to detoxify a wide range of drugs and hence their response to anaesthetics can differ considerably from adult animals.

When anaesthetizing neonates, it is essential to maintain body temperature using the techniques described in Chapter 4. Care must be taken to maintain good ventilation and to maintain fluid balance. In large species (dog, cat, sheep, pig) the umbilical vessels provide a convenient route for intravenous infusion.

It is preferable to use inhalational anaesthetics so that recovery is rapid and normal feeding is resumed as soon as possible. Methoxyflurane is particularly safe and effective in neonates. Neonatal animals usually require higher concentrations of anaesthetic, for example young adult rats require a concentration of approximately 2% halothane for maintenance of surgical anaesthesia, whereas neonates require 2–3%.

A. Anaesthesia using hypothermia

Deliberate induction of hypothermia has been used as a means of immobilizing neonatal rats and mice to enable surgical procedures to be undertaken. It is not clear whether the technique produces anaesthesia or simply immobility, although it seems likely that during the period of hypothermia the degree of depression of peripheral and central nervous system is sufficient to prevent the animal experiencing pain. It has been suggested, however, that recovery from hypothermia and the return of sensation to the body may be associated with pain, based on analogy with human experiences. No thorough investigations of the technique have been undertaken, and since safe and effective alternatives are available, for example methoxyflurane, halothane, fentanyl/droperidol (Park *et al.*, 1992) or fentanyl/fluanisone, it seems advisable to avoid the use of hypothermia whenever possible until its efficacy has been established.

6

Post-operative Care

Post-operative care must be considered a natural and essential extension of good anaesthetic technique. Failure to attend to the animal's needs during this critical period will inevitably delay recovery from anaesthesia. Poor post-operative care will exacerbate and prolong the metabolic disturbances caused by surgery, and if seriously neglected the animal may die.

All animals will require some degree of additional attention in the post-operative period and this is usually best achieved by providing a special recovery area. This will simplify the provision of the most appropriate environmental conditions, which will frequently differ from those present in a normal animal holding room. It will also highlight the special needs of animals placed in the recovery area and encourage extra attention from animal husbandry and nursing staff.

I. THE RECOVERY ROOM ENVIRONMENT

The recovery area for most laboratory mammals should be warm and quiet. Lighting should be subdued but adequate to allow easy observation of the animal. Higher intensity lighting must be readily available to enable more detailed examination and to allow procedures such as intravenous injection. In the immediate post-operative period, when homeothermic responses are depressed and the animal is still recovering from anaesthesia, the ambient temperature should be 27–30°C for adult animals, 35–37°C for neonates. Once the animal has recovered from the major depressant effects of the anaesthetic, the temperature can be reduced to 25°C for adults, but should be maintained at 35°C for neonates. This graduation in temperature can be achieved by maintaining a general room temperature of 21–25°C and providing supplemental heating using warming lamps or heating pads. Ideally, an animal incubator should be used: this will allow careful control of the ambient temperature and enable easy administration of oxygen. Although hypothermia is a potentially serious problem in the post-operative period, care must be taken not to overheat the animal and both the animal's rectal temperature and the temperature of its immediate environment should be carefully monitored. In small experimental units it may be impracticable to allocate space permanently for use as a recovery room. This should not lead to the concept being abandoned, as a temporary area within a laboratory can be set aside for this purpose. This can most easily be achieved by equipping a suitably sized trolley with an animal incubator and other necessary equipment. This can then be moved around the unit to wherever it is required.

A. Caging and bedding

In most instances small animals can be allowed to recover in their normal cages, placed either in a recovery room or inside an incubator (Thermocare, International Market Supply, Appendix 7). Small rodents and rabbits should not be allowed to recover from anaesthesia in cages that contain sawdust or wood shavings as bedding. This type of bedding will often stick to the animal's eyes, nose and mouth, so should be replaced by more suitable materials. A synthetic bedding with a texture similar to sheepskin (Vetbed, Cox Surgical, Appendix 7) has proved particularly useful for all species of animal. It is washable, autoclavable, extremely durable and appears to provide a comfortable surface for the animal. If such material is unavailable, towelling or a blanket should be used. Tissue paper is often provided as bedding for small rodents, but it is relatively ineffective as animals usually push it aside during recovery from anaesthesia and end up lying in the bottom of a plastic cage soiled with urine and faeces. Shredded paper of a type that will not stick to the animals' orifices or wounds should be used, since it provides a warm and comfortable nesting material (e.g. Paper shavings, RS Biotech*). Rabbits and guinea pigs should not be placed in grid-bottomed cages to recover from anaesthesia, but should be placed directly in an incubator or in a temporary plastic or cardboard holding box.

B. Nursing care

The response to human contact varies considerably among different species of animal and is influenced by previous experience. Excessive contact may have adverse, stressful effects in some small rodents and rabbits but most animals will benefit from some degree of nursing care carried out in a calm

* Tower Works, Well Street, Finedon, Northants NN9 5JP, UK;
Tel: 01933 680 133, Fax: 01933 680 155.

and reassuring manner. The degree of alarm caused to the animal can be reduced if it has been gradually familiarized to regular handling in the pre-operative period. This process forms an important part of pre-operative acclimatization in all species (see Chapter 1) and should be considered necessary when planning any series of experiments.

Most cats and dogs will respond positively to stroking and to a reassuring, familiar voice. If the recovery period is prolonged and normal grooming activity is not resumed, some animals, particularly dogs and cats, may respond favourably to regular grooming by nursing staff. Time should be given to encourage any dogs and cats who are reluctant to eat following surgery. Most can be tempted by hand-feeding. Warming the food will often make it more appetizing. Very often the presence of a familiar staff member to encourage eating will greatly affect appetite. Similar techniques can also be useful in pigs if they have been properly accustomed to human contact in the pre-operative period.

All species, including rodents and rabbits, should be checked at least twice a day. Particular attention should be given to cleaning the eyes, nose and mouth, which can become clogged with dried mucous or other debris. Monitoring of body weight and checking of wounds and surgical implants are also an important part of post-operative care. Rodents may be offered food pellets softened with warm water in bowls placed on the cage bottom, as many are often reluctant to reach up to food hoppers at this time.

It is important that a daily nursing routine is followed as far as possible. It is an advantage if one person is assigned to the care of post-surgical animals throughout the peri-operative period as they are more likely to notice any slight changes that may take place on a day-to-day basis. Careful record keeping is essential, so that other staff attending the animal, e.g. for out-of-hours emergencies, will be aware of all treatments given and the animal's progress.

II. PROBLEMS DURING THE RECOVERY PERIOD

A. Vomiting, regurgitation and hypostatic pneumonia

The swallowing and cough reflexes are usually suppressed during anaesthesia and these gradually return as anaesthesia lightens. If an endotracheal tube is present it should be removed when the animal begins to swallow spontaneously or attempts to cough. Non-ruminant species should be placed on their sides, with head and neck extended, to try to minimize the risk of airway obstruction. If the animal is recumbent for more than 4 hours then it should be repositioned to lie on its other side to prevent passive congestion of the lungs and the development of hypostatic pneumonia. In large animals such as dogs and farm animals it may be necessary to massage areas such as the elbow and hock, to prevent pressure sores developing. If prolonged recumbency is anticipated, it may be advisable to protect these areas with padded bandages.

If the animal begins to vomit, an attempt should be made to position it so that its head is below the level of the thorax and abdomen to try to prevent aspiration of the vomit. If practicable, the mouth and pharynx should be cleared using a vacuum suction, or a piece of suitably sized tubing attached to a 50 ml syringe. Oxygen should be administered and, if inhalation of vomit is believed to have occurred, corticosteroids should be administered (30 mg kg^{-1} i/v of methylprednisolone), together with a broad-spectrum antibiotic.

Ruminants (sheep, goats and cattle) can present particular problems during recovery from anaesthesia. They should be propped up on their sternums to minimize the risk of overdistension of the rumen with gas (ruminal tympany) and of inhalation of regurgitated rumen contents. If ruminal tympany develops, it should be relieved immediately either by passing a stomach tube or by puncturing the rumen through the body wall with a large-bore trocar. If a trocar is not available, the largest possible needle (preferably 12 gauge or larger) should be used. If the member of nursing staff concerned is not familiar with this technique, veterinary advice should be sought immediately.

B. Respiratory depression

The respiratory depression produced by most anaesthetic agents frequently persists into the post-operative period. The degree of depression may also increase post-operatively and this may go unnoticed until severe hypercapnia and hypoxia have developed. It is preferable to continue to monitor the respiratory system, and a pulse oximeter is ideal for this purpose in larger species (see Chapter 4). If not already in use, the probe can be attached to the animal in theatre, and a battery-operated instrument used to monitor the animal during movement to the recovery area. The probe can be left taped in place on the tail or on a digit until the animal has regained its righting reflex. If a pulse oximeter is unavailable, then other forms of respiratory monitor can be used, for example positioning a thermistor close to the animal's nose (IMP Respiratory Monitor, Veterinary Instrumentation, Appendix 7). At the very least, regular clinical observation of the animal should be made and the respiratory rate recorded. If respiratory depression is noted, it should be treated using a respiratory stimulant such as doxapram and by the administration of oxygen. As doxapram has a relatively short duration of action (10–15 minutes), it may be necessary to administer repeated doses or to establish a continuous infusion of the drug.

Many animals appear to benefit if oxygen administration is continued into the post-operative period. This is best achieved in small animals by piping the gas into an animal incubator, but in large animals it is often more practicable to tape a small, soft-ended catheter at the external nares and use this to administer the gas.

C. Fluid therapy

The voluntary water intake of all animals should be recorded post-operatively, even if this consists simply of making a rough estimate based on the level remaining in a water bottle. Fluid intake is frequently reduced post-operatively and if dehydration is allowed to develop, it can seriously compromize the recovery of the animal. Fluid requirements of most species are approximately 40–80 ml kg^{-1} every 24 hours, but vomiting, diarrhoea or other abnormal losses will increase this requirement.

If the animal is fully conscious, supplemental fluid is best given by the oral route. If the animal is unable or unwilling to accept oral administration,

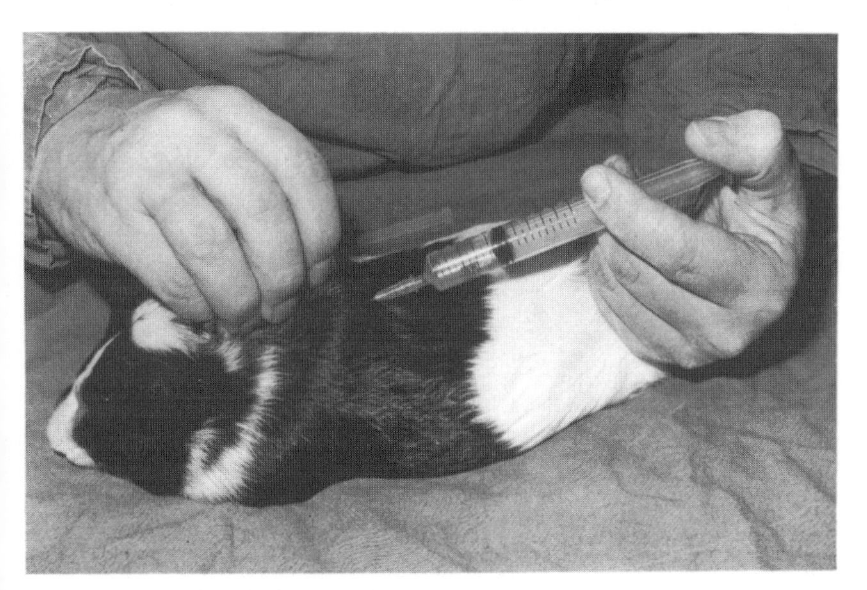

Fig. 6.1. Subcutaneous administration of fluid to the guinea pig immediately prior to recovery from anaesthesia.

TABLE 6.1
Approximate volumes for fluid replacement therapy by intraperitoneal or subcutaneous administration

	Subcutaneous (ml)	Intraperitoneal (ml)
Cat (3 kg)	50	50–100
Gerbil (60 g)	1–2	2–3
Guinea pig (1 kg)	10–20	20
Hamster (100 g)	3	3
Marmoset (500 g)	5–10	10–15
Mouse (30 g)	1–2	2
Rabbit (3 kg)	30–50	50
Rat (200 g)	5	5

then dextrose-saline (4% dextrose, 0.18% saline) or saline (0.9%) can be given quickly and easily by the subcutaneous or intraperitoneal route (Figure 6.1, Table 6.1). Severe dehydration causes loss of skin tone that causes it to tent and tend to remain elevated when a fold is twisted between the fingers. In larger animals, dehydration will result in the mucous membranes becoming dry to the touch. If this degree of dehydration has inadvertently been allowed to develop, fluids must be administered intravenously.

The monitoring of body weight in the pre-operative and post-operative periods can provide a good indication of the adequacy of fluid intake. Although a small fall in body weight will be recorded because of the almost inevitable reduction in food intake that occurs post-operatively, most weight loss represents a fluid deficit.

Along with assessment of food and water intake, the urinary and faecal output of the animal should be recorded and any abnormalities investigated. As with most of these variables, a meaningful judgment can only be made if the animal has been observed in the pre-operative period. A reduction in urine output may be the result of dehydration, urinary tract injury or the animal suffering pain. If the bladder is full it may require catheterization to empty it. This is a relatively simple technique, but requires some degree of expertise and it will usually be preferable to consult a veterinary surgeon or experienced animal technician. If catheterization is not possible, it may prove necessary to drain the bladder by direct puncture through the body wall. This procedure should only be attempted by individuals who have undergone training in the technique.

If the animal fails to pass faeces, this may be due simply to an absence of faecal material because of pre-operative fasting. It may also be caused by a paralytic ileus, due to excessive handling of the bowel during a laparotomy. In some instances the animal may be constipated and require administration of an enema (e.g. Microlax, SmithKline Beecham). Defaecation may also be suppressed if the animal is in pain, particularly following a laparotomy.

It is important that these and the other observations described above are recorded carefully. It is helpful to provide a standard record card for each animal, which will encourage nursing staff to complete the required observations. It will also allow easy and rapid reference by staff who may be called in to deal with any problems that might arise.

C. Prevention of wound infection

Provided that careful aseptic surgical techniques have been employed, it may be considered unnecessary to administer antibiotics routinely to animals in the post-operative period. In addition, some species appear to show a remarkable resistance to the development of wound sepsis and appear to tolerate standards of cleanliness that would be totally unacceptable in human medical practice. This apparent resistance to infection must not be used as an excuse for poor surgical standards and every effort should be made to adopt aseptic techniques for all animal surgical procedures. It has been demonstrated, for example, that rats are not only susceptible to infection, but also show behavioural changes following establishment of wound infections (Bradfield *et al.*, 1992). It is therefore important that all species of animal should be monitored carefully for any signs of infection (Morris, 1995).

Since animals will almost inevitably soil their wounds with faeces and urine, administration of prophylactic antibiotics may be useful in minimizing the risk of infection. One problem of providing peri-operative treatment with antibacterials is the risk of inducing enterotoxaemia in some species, particularly the guinea pig, hamster and rabbit. The use of antibacterial agents in rodents and rabbits has been reviewed by Morris (1995), and provided that care is taken in the choice of agent, such problems can be avoided. Suggested dose rates of antibiotics for each species are given in Table 6.2.

TABLE 6.2a

Antibiotic and antibacterial drug doses for laboratory animals. Note that the majority of these doses are based solely on clinical experience, since pharmacokinetic data are not available for these species. Before administering any of these compounds, research workers are strongly advised to consult their laboratory animal veterinarian for advice on drug selection and duration of treatment. For a comprehensive review of the effects of antibiotics in laboratory species, see Morris (1995)

	Mouse	Rat	Hamster	Gerbil	Guinea pig	Rabbit
Cephalexin	15 mg kg^{-1} i/m b.i.d.	15 mg kg^{-1} s/c b.i.d.	–	25 mg kg^{-1} s/c u.i.d.	15 mg kg^{-1} s/c u.i.d.	15 mg kg^{-1} s/c b.i.d.
Chloramphenicol	50 mg kg^{-1} s/c b.i.d.	10 mg kg^{-1} i/m b.i.d.	30 mg kg^{-1} s/c b.i.d.	30 mg kg^{-1} s/c b.i.d.	20 mg kg^{-1} i/m b.i.d.	15 mg kg^{-1} i/m b.i.d.
Enrofloxacin	8.5 mg kg^{-1} s/c b.i.d.	10 mg kg^{-1} s/c b.i.d.	10 mg kg^{-1} s/c b.i.d.	10 mg kg^{-1} s/c b.i.d.	5–10 mg kg^{-1} s/c b.i.d.	5–10 mg kg^{-1} s/c b.i.d.
Neomycin	2 mg ml^{-1} in drinking water	2 mg ml^{-1} in drinking water	250 mg kg^{-1} per os in divided doses	100 mg kg^{-1} per os in divided doses	5 mg kg^{-1} per os b.i.d.	0.2–0.8 mg ml^{-1} drinking water
Co-trimazine 40/200	30–50 mg kg^{-1} s/c b.i.d.	30–50 mg kg^{-1} s/c b.i.d.	30 mg kg^{-1} s/c b.i.d.	30 mg kg^{-1} s/c b.i.d.	30–50 mg kg^{-1} s/c b.i.d.	30–50 mg kg^{-1} s/c b.i.d.
Tylosin	–	10 mg kg^{-1} s/c u.i.d.	10 mg kg^{-1} s/c u.i.d.	10 mg kg^{-1} s/c u.i.d.	–	–

TABLE 6.2b

Antibiotic and antibacterial drug doses for laboratory animals. Note that the majority of these doses for ferrets and non-human primates are based solely on clinical experience, since pharmacokinetic data are not available for these species. Before administering any of these compounds, research workers are strongly advised to consult their laboratory animal veterinarian for advice on drug selection and duration of treatment. For a comprehensive review of the effects of antibiotics in laboratory species, see Morris (1995)

	Ferret	Cat	Dog	Pig	Sheep	Primate
Amoxycillin	7 mg kg^{-1} s/c u.i.d.	7 mg kg^{-1} s/c u.i.d.	7 mg kg^{-1} s/c u.i.d.	7 mg kg^{-1} i/m u.i.d.	7 mg kg^{-1} i/m u.i.d.	7 mg kg^{-1} s/c u.i.d.
Cephalexin	10 mg kg^{-1} s/c u.i.d.	10 mg kg^{-1} s/c u.i.d.	10 mg kg^{-1} s/c u.i.d.	10 mg kg^{-1} i/m u.i.d.	10 mg kg^{-1} i/m u.i.d.	–
Chloramphenicol	25 mg kg^{-1} s/c u.i.d.	25 mg kg^{-1} s/c u.i.d.	50 mg kg^{-1} s/c u.i.d.	11 mg kg^{-1} i/m u.i.d.	–	20 mg kg^{-1} i/m b.i.d.
Enrofloxacin	–	5 mg kg^{-1} s/c u.i.d.	5 mg kg^{-1} s/c u.i.d.	–	–	5 mg kg^{-1} s/c b.i.d.
Neomycin	10 mg kg^{-1} per os u.i.d. in divided doses	10 mg kg^{-1} per os u.i.d. in divided doses	10 mg kg^{-1} per os u.i.d. in divided doses	11 mg kg^{-1} per os b.i.d.	11 mg kg^{-1} per os b.i.d.	10 mg kg^{-1} per os b.i.d.
Tylosin	–	20–45 mg kg^{-1} per os u.i.d. in divided doses	2–10 mg kg^{-1} i/m u.i.d.	2–10 mg kg^{-1} i/m u.i.d.	–	–

III. MANAGEMENT OF POST-OPERATIVE PAIN

The effective alleviation of post-operative pain in laboratory animals should be considered an important goal in all research establishments. Despite the emphasis given to humane treatment of laboratory animals in the national legislation of many countries, analgesics may still not be administered routinely in the post-operative period. This omission is particularly common when the animals concerned are small rodents. When analgesics are administered, assessment of their efficacy is usually based on highly subjective criteria. The lack of an objective means of pain assessment may account in part for the relatively infrequent use of analgesics in animals, in comparison with their use in humans. This is not meant to imply that veterinary surgeons and others involved in animal care are incapable of recognizing that an animal is in pain, but preconceptions about animal pain may limit the value of any assessment of its severity (see 'Pain assessment' below).

Human pain has been defined as 'an unpleasant sensory and emotional experience associated with actual or potential tissue damage or described in terms of such damage' (IASP, 1979). Our concern to alleviate pain in animals is based upon extrapolation from human experience, and it seems reasonable to assume that the sensory experiences of pain are present in animals. The mechanisms of nociception in animals have been extensively studied (Short and Van Poznak, 1992) and found to resemble closely those present in humans. The central pathways presumed necessary for the emotional component of pain are also present. Nevertheless, the demonstration of equivalent anatomical structures and physiological processes does not provide conclusive evidence that both the sensory and emotional components of the experience of pain are similar in animals and humans. It is possible that the relative significance, magnitude and duration of pain in response to particular types of injury may all vary in animals. This reasoning can lead to a downgrading of the significance of pain in animals and can be used as an explanation for the apparent rapid recovery of animals from surgical procedures that would be incapacitating in humans. It also leads to a belief that pain relief is unnecessary or unimportant in animals. Since our current knowledge of the significance of pain in animals is severely limited, it is best to assume that pain is an unpleasant experience in animals, that it is an experience they would normally avoid, and that it is our responsibility to alleviate pain whenever possible.

Although we would wish to alleviate pain because of concerns for animal welfare, a number of counter-arguments have been advanced to justify withholding analgesics. It has been claimed, for example, that alleviation of post-operative pain will result in the animal injuring itself. Pain, of course, does have a protective function and is of value in warning of tissue damage in an individual. Pain arising from injured tissues often results in the animal or human immobilizing the affected area, which helps to prevent further injury. Nevertheless, pain is also harmful since the immobility and muscle spasm it produces can cause muscle wasting and weakness. Thoracic and abdominal pain may reduce ventilation and cause hypoxia and hypercapnia. Pain may also cause a marked reduction in food and water consumption (Liles and Flecknell, 1993a,b). Pain in humans has been shown to prolong the metabolic response to surgery (Kehlet, 1978), to increase the requirement for hospital care following operative procedures (Alexander and Hill, 1987) and to have a range of other detrimental effects (Breivik, 1994). Provided that surgery has been carried out competently, administration of analgesics, which allow resumption of normal activity, rarely results in problems associated with the removal of pain's protective function. Claims that analgesic administration results in skin suture removal are unsubstantiated and contrary to findings in the author's laboratory. In certain circumstances, for example after major orthopaedic surgery, additional measures to protect and support the operative site may be required, but this is preferable to allowing an animal to experience unrelieved pain. All that is required in these circumstances is a temporary reduction in the animal's cage or pen size, or the provision of additional external fixation or support for the wound. It must be emphasized that these measures are very rarely necessary, and in the author's institute administration of analgesics to laboratory animals after a wide variety of surgical procedures has not resulted in any adverse clinical effects.

In human clinical practice, analgesic drugs are frequently withheld because of fears of their undesirable side-effects such as respiratory depression and addiction (Smith, 1984). The side-effects of opiates in animals are generally less marked than in humans and should rarely be a significant consideration when planning a post-operative care regimen. Another factor that may limit the use of analgesic drugs is a lack of knowledge of appropriate dose rates and dosage regimens. This is primarily a problem of poor dissemination of existing information. Virtually every available analgesic drug has undergone extensive testing in animals. Dose rates are therefore available for a range of drugs in many common laboratory species (Flecknell, 1984; Liles and Flecknell, 1992). It is occasionally difficult to extrapolate available dose rates from one species to another and to translate dose rates that are effective in experimental analgesiometry into dose rates that are appropriate for clinical use. Nevertheless, in most instances a reasonable guide as to a suitable, and safe, dose rate can be obtained.

It has also been suggested that many research scientists are reluctant to administer pain-relieving drugs because their use might adversely affect the results of an experiment. Although there will be occasions when the use of one or other type of analgesic is contraindicated, it is extremely unlikely that there will be no suitable analgesic that could be administered. More usually, the reluctance to administer analgesics is based upon the misconceived idea that the use of any additional medication in an experimental animal is undesirable. The influence of analgesic administration in a research protocol should be considered in the context of the overall response of the animal to anaesthesia and surgery. As discussed in Chapter 3, the responses to surgical stress may overshadow any possible adverse interactions associated with analgesic administration. An additional consideration is that many arrangements for intra-operative care fail to control variables such as body temperature, respiratory function and blood pressure. It seems illogical to assume that changes in the function of the cardiovascular or respiratory systems are unimportant, but that administration of an analgesic will be of overriding significance. It should be considered an ethical responsibility of a research worker to provide a reasoned, scientific justification if analgesic drugs are to be withheld. It is also important to realize that the presence of pain can produce a range of undesirable physiological changes, which may radically alter the rate of recovery from surgical procedures (Keeri-Szanto, 1983).

A. Assessment of pain

The approach to animal pain based on comparative biology, outlined at the start of this section, leads naturally to the assumption that conditions that would cause pain in humans would also lead to the production of pain in animals. When examining such animals, we interpret certain clinical signs as suggesting the presence of pain. Following a surgical procedure, a dog might howl or whimper, perhaps guarding the surgical wound, and show signs of avoidance or aggression when the area is handled. These types of behaviour are easy to equate with the behaviour of humans in pain, so we readily diagnose animals showing these clinical signs as being in pain and may then give analgesics. Unfortunately, this anthropomorphic view of pain is flawed. Many animals do not respond to conditions and procedures that would cause human pain in a way that is immediately apparent as pain-related behaviour. For example, following routine laparotomy, rats do not show obvious signs of pain. This is an apparent contradiction of the previous hypothesis based on comparative biology — human patients do experience significant pain after abdominal surgery, most complain about their pain, and most require opioids or other forms of analgesia. This discrepancy between the actual behaviour of animals and the behaviour that would be predicted from human experience gives rise to the view that although pain may occur in animals, it is less severe than in humans, and that the more resilient nature of animals results in more rapid recovery with less experience of pain. The natural consequence of this is to assume that animals do not require analgesics as frequently as human patients, and perhaps may not even require them at all. The key to introducing effective pain control may therefore be to improve our methods of pain recognition and assessment. For example, in the case of laparotomy mentioned earlier, if the animal is observed for a prolonged period and its behaviour analysed carefully, then more subtle changes become apparent (Liles, 1994). These changes may be normalized by administration of an analgesic, and this supports the view that they may be related to the presence of post-operative pain.

Pain assessment is important not simply because it would encourage greater use of analgesics, but because it would also encourage more appropriate use of these drugs. In many research animal units, national legislation requires that pain should be assessed, based on the assumption that procedures that are painful in humans will be equally painful in animals. Following a surgical procedure, it is therefore assumed that an analgesic will be required. The choice of analgesic should be determined in some part by the degree of pain that is present, since inappropriate use of potent analgesics may lead to the undesirable side-effects of these agents outweighing any benefits arising from alleviation of pain. Similarly, use of low-potency agents in circumstances in which severe pain is present will result in insufficient pain relief. The simple assumption that after identical surgical procedures, the degrees of pain present in all species of animal and in humans will be identical is highly unlikely to be correct. Even if this broad comparison were possible, it would also be necessary to assume that the duration of pain, and hence requirement for pain relief, was identical in all species including man in equivalent circumstances. It also fails to account for individual variation in response to analgesics. Following identical surgical procedures, different human patients can have markedly different analgesic requirements (Alexander and Hill, 1987). This has become clearly apparent with the introduction of patient-controlled analgesia, which removes some of the obstacles to analgesic administration (Skues *et al.*, 1993). Although equivalent data in laboratory animals are limited to anecdote and clinical opinion, data from analgesiometry in research animals show similar variation, both among individuals (Cowan *et al.*, 1977a,b) and between animals of different inbred strains (Pick *et al.*, 1991). In addition, both age and sex have been shown to influence responses to analgesics (Frommel and Joye, 1964). Selection of an arbitrary dose of analgesic will therefore almost certainly lead to overdosage of some animals, and provision of inadequate analgesia for others. Development of reliable methods of pain assessment would enable analgesics treatment to be tailored to suit the needs of each individual animal.

A number of different approaches to pain assessment in animals have been suggested. One of the most influential papers was that published by Morton and Griffiths (1985), which advocated the use of a scoring system based on a wide range of clinical signs. This paper influenced a large number of other groups, who modified the original hypothesis but retained the central notion of identifying pain-specific behaviours, and rating them in some way (Association of Veterinary Teachers and Research Workers (AVTRW), 1986; LASA, 1990; Flecknell, 1991; ILAR, 1992; FELASA, 1994). Unfortunately, progress in validating this hypothesis has been limited. An early report (Leese *et al.*, 1988) showed that the technique could be applied successfully, but the few subsequently published data have been less encouraging. Particular problems noted were the considerable between-observer variation and the poor predictive value of certain of the parameters scored (Beynen *et al.*, 1987, 1988). The between-observer variation is not unexpected, and parallels problems recognized in human pain scoring. It appears that if the number of observers is restricted, and the criteria used carefully selected, reasonable agreement can be achieved (Reid and Nolan, 1991). Although problems in pain scoring have been encountered, the value of behavioural observations has been supported by studies in companion and farm animals, for example the effects of tail docking and castration in lambs (Wood *et al.*, 1991) and castration in piglets (McGlone and Hellman, 1988). Other studies have examined behavioural changes, food and water consumption and body weight as potential indices of pain. The results of these studies are encouraging (Flecknell and Liles, 1991, 1992; Liles and Flecknell, 1993a, b; Liles, 1994), but suggest that the use of behavioural monitoring in laboratory species may be difficult unless animals are in moderate or severe pain.

Although well-validated quantitative methods of post-operative pain assessment have yet to be developed, virtually all studies of post-operative pain in animals have demonstrated a beneficial effect of analgesic therapy (Flecknell, 1994). It is not unreasonable, then, to suggest that most animals require some analgesics post-operatively. It is recommended that attempts are made to assess pain using a modification of the Morton and Griffiths technique, by selecting the variables used for scoring to suit the specific animal model concerned. The choice of variable is made after observing a small number of animals following surgery, and determining which variables are most affected by the procedure. The types of changes that may be assessed are outlined below.

1. Activity

As mentioned above, the overall level of activity of an animal suffering pain is usually reduced and most laboratory species will tend to remain motionless in a corner of their cage. Occasionally, an animal may show unusual restlessness and may appear unable to relax. When the animal does move, its posture or gait may be altered. This is most obviously seen when limb pain is present, but is often noted following laparotomy, when the back may be arched to reduce tension on the abdominal muscles. This altered posture, coupled with a tendency to shorten the length of each stride, can be seen both in rodents and rabbits, and also dogs, cats and farm animals. Pain from an abdominal incision may also affect the frequency of urination and defaecation in species in which this process requires marked abdominal muscle contraction. Particular behaviours such as climbing, rearing up on the hindlimbs, stretching and scratching may also be affected, but careful observation by an experienced assessor may be necessary before such changes are noticed. A further complication in assessing behaviour is that the animal may change its responses in the presence of an unfamiliar observer. In addition, some species are nocturnal, and observation of normal behaviour will require attendance during the dark phase of its photoperiod. Both of these problems can be resolved to some extent by using video cameras to monitor the animal's behaviour.

2. Appearance

Even when at rest, the animal's overall appearance may be altered. The animal may adopt a hunched-up posture and position itself in a corner of its cage or pen. Pain may result in a reduction in grooming activity, which leads to the development of an unkempt appearance of the coat and soiling of the anus. Lack of grooming may also lead to the build-up of an encrusted discharge around the eyes, nose and mouth. Rats may develop dark encrustations around the eyes or nose. This material is porphryin, excreted from the harderian glands, and if wiped with moist cotton wool it has a red colour. The presence of porphyrin staining is a non-specific stress response, but should alert the observer to the possibility that the stress involved may be pain.

3. Temperament

Changes in temperament often occur in animals experiencing pain. Previously tractable animals may become uncharacteristically aggressive and may bite or scratch. Alternatively, a previously active animal which showed obvious interest in its handler may appear completely apathetic. The animal may cower away from the handler and attempt to avoid being restrained. The interpretation of any of these types of behaviour will require a knowledge not only of the normal predicted behaviour of an animal of that particular age, sex and species, but also prior knowledge of the normal behaviour of that particular individual. Clearly, close liaison with animal care staff is essential in attempting to assess the behaviour of an animal in the post-operative period.

4. Vocalizations

Acute pain can make an animal cry out and handling an animal which is in pain may provoke such a response. The pitch of the cry may be abnormal and may be accompanied by attempts to bite the handler or to escape. Animals in pain rarely cry continuously, although on occasions dogs may howl or whimper for long periods and prolonged vocalization may also be made by sheep and cattle. When assessing pain in rodents, it is important to appreciate that many of their cries are at high sound frequencies which are inaudible to humans.

5. Feeding behaviour

Food and water intake are often markedly reduced if the animal is in pain. Severe pain is often associated with a complete cessation of eating and drinking. These changes in feeding may go unnoticed if the animal is fed ad lib from a hopper, or if other animals that are feeding normally are present in the cage or pen. A reduction in body weight as a consequence of this inappetence can usually be readily detected, but normal day-to-day variations in body weight must also be appreciated. To improve the detection of changes in food and water intake, weighed quantities of food and water should be dispensed and daily intake measured. Weighing the food hopper and water bottle provides a satisfactory means of monitoring intake in larger rodents. Care must be taken that spillage of food by the animal does not result in intake being erroneously assessed as normal. In addition to recording food consumption, the animal should be weighed each day to determine any changes in body weight.

A reduction in food and water intake will also be reflected in a reduction in faecal and urine output, but the latter may be difficult to detect. The onset of dehydration will be reflected in the clinical appearance of the animal. Loss of skin tone will cause it to tent and tend to remain elevated when a fold is twisted between the fingers. In larger animals, dehydration will result in the mucous membranes becoming dry to the touch.

6. Alterations in physiological variables

Pain generally causes changes in the pattern and rate of respiration. This can be dramatic following thoracic surgery, when the reduction in depth of respiration can cause considerable concern. In other instances the change may be less obvious and may be masked by the normal tendency of animals such as rodents or rabbits to respond to restraint or close observation with an increase in their respiratory rate.

Pain may also affect the cardiovascular system. Frequently, heart rate is increased, but the natural responses to handling may mask these changes. Severe pain may cause the development of circulatory failure (shock), with blanching and chilling of the extremities and a decrease in the strength of the peripheral pulse.

B. Pain relief

Leaving aside the problems of pain assessment, empirical treatment of presumed painful conditions will continue, and it is not unreasonable to assume that analgesic therapies shown to be effective in humans are likely also to be effective in animals. Although the assessment of clinical efficacy may not have been completed, studies of novel analgesic compounds and delivery systems in animals have established their safety and efficacy in analgesiometric tests. Analgesics can be broadly divided into two groups: the opioids or narcotic analgesics, and the non-steroidal anti-inflammatory drugs (NSAIDs) such as aspirin. Local anaesthetics can also be used to provide post-operative pain relief by blocking all sensation from the affected area. Suggested dose rates of analgesics are given in Table 6.3.

1. NSAIDs

Traditionally, NSAIDs have been considered as low-potency analgesics, suitable for the control of mild pain, or as agents primarily for use in conditions such as arthritis, where the inflammatory component of the disease process was responsible for some or all of the pain. The perception of NSAIDs has changed with the introduction of a number of compounds

TABLE 6.3a
Suggested dose rates for non-steroidal anti-inflammatory drugs in laboratory animals

Drug	Mouse	Rat	Guinea pig	Rabbit	Ferret
Aspirin	120 mg kg^{-1} per os	100 mg kg^{-1} per os	87 mg kg^{-1} per os	100 mg kg^{-1} per os	200 mg kg^{-1} per os
Carprofen	–	5 mg kg^{-1} s/c	–	1.5 mg kg^{-1} per os b.i.d.	–
Diclofenac	8 mg kg^{-1} per os	10 mg kg^{-1} per os	2.1 mg kg^{-1} per os	–	–
Flunixin	2.5 mg kg^{-1} s/c, i/m, ? 12 hourly	2.5 mg kg^{-1} s/c, i/m, ? 12 hourly	–	1.1 mg kg^{-1} s/c, i/m, ? 12 hourly	0.5–2 mg kg^{-1} s/c, 12–24 hourly
Ibuprofen	30 mg kg^{-1} per os	15 mg kg^{-1} per os	10 mg kg^{-1} i/m ? 4 hourly	10 mg kg^{-1} i/v, ? 4 hourly	–
Indomethacin	1 mg kg^{-1} per os	2 mg kg^{-1} per os	8 mg kg^{-1} per os	12.5 mg kg^{-1} per os	–
Ketoprofen	–	–	–	3 mg kg^{-1} i/m	–
Paracetamol (acetominophen)	200 mg kg^{-1} per os	200 mg kg^{-1} per os	–	–	–
Piroxicam	3 mg kg^{-1} per os	3 mg kg^{-1} per os	6 mg kg^{-1} per os	–	–

Note that considerable individual and strain variation in response may be encountered and that it is therefore essential to assess the analgesic effect in each individual animal.

TABLE 6.3b
Suggested dose rates for non-steroidal anti-inflammatory drugs in laboratory animals

Drug	Pig	Sheep	Primate	Dog	Cat
Aspirin		50–100 mg kg^{-1} per os, 6–12 hourly	20 mg kg^{-1} per os, 6–8 hourly	10–25 mg kg^{-1} per os, 8 hourly	10–25 mg kg^{-1} per os, every 48 hours
Carprofen	2–4 mg kg^{-1} i/v or s/c, once daily	?	?	4 mg kg^{-1} i/v or s/c, once daily 1–2 mg kg^{-1} b.i.d. per os, for 7 days	4 mg kg^{-1} s/c or i/v
Flunixin	1–2 mg kg^{-1} i/v or s/c, once daily	2 mg kg^{-1} i/v or s/c, once daily	2–4 mg kg^{-1} s/c, u.i.d.	1 mg kg^{-1} i/v or i/m, 12 hourly 1 mg kg^{-1} per os, daily for up to 3 days	1 mg kg^{-1} s/c, daily for up to 5 days
Ibuprofen				10 mg kg^{-1} per os, 24 hourly	
Ketoprofen				2 mg kg^{-1} s/c, i/m or i/v, once daily for up to 3 days 1 mg kg^{-1} per os, daily for 5 days	1 mg kg^{-1} s/c, once daily for up to 3 days 1 mg kg^{-1} per os, once daily for up to 5 days
Paracetamol (acetominophen)				15 mg kg^{-1} per os, 6–8 hourly	Contraindicated
Piroxicam				300 μg kg^{-1} every 48 hours	

Note that considerable individual and strain variation in response may be encountered, and that it is therefore essential to assess the analgesic effect in each individual animal.

TABLE 6.3c

Suggested dose rate for opioid analgesics in laboratory animals

Drug	Mouse	Rat	Guinea pig	Rabbit	Ferret
Buprenorphine	0.05–0.1 mg kg^{-1} s/c, 12 hourly	0.01–0.05 mg kg^{-1} s/c or i/v, 8–12 hourly 0.1–0.25 mg kg^{-1} per os, 8–12 hourly	0.05 mg kg^{-1} s/c, 8–12 hourly	0.01–0.05 mg kg^{-1} s/c or i/v, 8–12 hourly	0.01–0.03 mg kg^{-1} i/v, i/m or s/c, 8–12 hourly
Butorphanol	1–5 mg kg^{-1} s/c, 4 hourly	2 mg kg^{-1} s/c, 4 hourly	–	0.1–0.5 mg kg^{-1} i/v, 4 hourly	0.4 mg kg^{-1} i/m, 4–6 hourly
Morphine	2.5 mg kg^{-1} s/c, 2–4 hourly	2.5 mg kg^{-1} s/c, 2–4 hourly	2–5 mg kg^{-1} s/c or i/m, 4 hourly	2–5 mg kg^{-1} s/c or i/m, 2–4 hourly	0.5–5 mg kg^{-1} i/m or s/c, 6 hourly
Nalbuphine	4–8 mg kg^{-1} i/m, ? 4 hourly	1–2 mg kg^{-1} i/m, 3 hourly	1–2 mg kg^{-1} i/v, i/p or i/m	1–2 mg kg^{-1} i/v, 4–5 hourly	–
Pentazocine	10 mg kg^{-1} s/c, 3–4 hourly	10 mg kg^{-1} s/c, 3–4 hourly		5–10 mg kg^{-1} s/c or i/m, i/v, 4 hourly	–
Pethidine (meperidine)	10–20 mg kg^{-1} s/c or i/m, 2–3 hourly	10–20 mg kg^{-1} s/c or i/m, 2–3 hourly	10–20 mg kg^{-1} s/c or i/m, 2–3 hourly	10 mg kg^{-1} s/c or i/m, 2–3 hourly	5–10 mg kg^{-1} i/m or s/c, 2–4 hourly

Note that considerable individual and strain variation in response may be encountered, and that it is therefore essential to assess the analgesic effect in each individual animal.

TABLE 6.3d
Suggested dose rates for opioid analgesics in laboratory animals

Drug	Pig	Sheep	Primate	Dog	Cat
Buprenorphine	0.005–0.02mg kg^{-1} i/m or i/v, 6–12 hourly	0.005–0.01 mg kg^{-1} i/m or i/v, 4 hourly	0.005–0.01 mg kg^{-1} i/m or i/v, 6–12 hourly	0.005–0.02 mg kg^{-1} i/m, s/c or i/v, 6–12 hourly	0.005–0.01 mg kg^{-1} s/c or i/v, 8–12 hourly
Butorphanol	–	–	0.01 mg kg^{-1} i/v, ? 3–4 hourly	0.2–0.4 mg kg^{-1} s/c or i/m, 3–4 hourly	0.4 mg kg^{-1} s/c, 3–4 hourly
Morphine	0.2–1 mg kg^{-1} i/m, ? 4 hourly	0.2–0.5 mg kg^{-1} i/m, ? 4 hourly	1–2 mg kg^{-1} s/c or i/m, 4 hourly	0.5–5 mg kg^{-1} s/c or i/m, 4 hourly	0.1 mg kg^{-1} s/c, 4 hourly
Nalbuphine	–	–	–	0.5–2.0 mg kg^{-1} s/c or i/m, 3–4 hourly	1.5–3.0 mg kg^{-1} i/v, 3 hourly
Oxymorphone	–	–	–	0.05–0.22 mg kg^{-1} i/m, s/c or i/v, 2–4 hourly	0.2 mg kg^{-1} s/c or i/v
Pentazocine	2 mg kg^{-1} i/m or i/v, 4 hourly	–	2–5 mg kg^{-1} i/m or i/v, 4 hourly	2 mg kg^{-1} i/m or i/v, 4 hourly	–
Pethidine (meperidine)	2 mg kg^{-1} i/m or i/v, 2–4 hourly	2 mg kg^{-1} i/m or i/v, 2–4 hourly	2–4 mg kg^{-1} i/m or i/v, 2–4 hourly	10 mg kg^{-1} i/m, 2–3 hourly	2–10 mg kg^{-1} s/c or i/m, 2–3 hourly

Note that considerable individual and strain variation in responses may be encountered, and that it is therefore essential to assess the analgesic effect in each individaul animal.

that have been shown to have considerable analgesic potency (Cunningham and Lees, 1994). In laboratory species, data from a number of analgesiometric tests provide a basis for estimating appropriate dose rates for clinical use in these species (Liles and Flecknell, 1992). Estimating frequency of administration is much more difficult, however, since there are considerable variations in elimination times for NSAIDs in different species (DeBuf, 1994; Lees *et al.*, 1991). Despite these problems, there are now a range of NSAIDs with clear indications for use in alleviating pain in animals (DeBuf, 1994). The most significant problems associated with NSAID administration are gastrointestinal disturbances, notably ulceration and haemorrhage, nephrotoxicity, and interference with platelet function. Other blood dyscrasias and liver toxicity can also occur (Lees *et al.*, 1991; Liles and Flecknell, 1992). These side-effects are seen primarily following prolonged administration, and are rarely of significance when treatment is for 2–3 days post-operatively. It should be noted that some NSAIDs (e.g. aspirin) have been reported as causing fetal abnormalities, so their use should be avoided in pregnant animals. In the research environment, the non-specific effects of NSAIDs may preclude their administration in certain research protocols. However, they offer an alternative to opioids, which have different non-specific side-effects. Consideration of the nature of the research study and potential interactions with analgesics allows a logical choice of analgesic agent to be made.

a. Aspirin. Aspirin can be used to alleviate mild pain. It is most effective in humans for musculoskeletal pain, and is less effective for visceral pain. Dispersible and enteric-coated tablets are available. Injectable formulations are generally available only for research use. A wide range of preparations that combine aspirin with other analgesics — e.g. paracetamol (acetominophen), codeine and dextropropoxyphene — are available for human use, but their efficacy in animals has not been evaluated.

b. Paracetamol (acetominophen). Paracetamol has similar analgesic efficacy to aspirin but has little anti-inflammatory activity. It causes less gastrointestinal irritation, but overdosage causes liver toxicity. These analgesics should not be administered to cats because of problems of toxicity. Tablets and oral suspensions are available for human use, and these may be used in a wide range of laboratory species, although few data concerning efficacy in animals are available (Mburu *et al.*, 1988).

c. Ibuprofen. Ibuprofen is effective against mild pain in humans, but clinical trials in animals have not been undertaken. Both tablets and suspensions are available for human use.

d. Phenylbutazone. Phenylbutazone has been widely used for controlling mild pain in larger species. Analgesiometric studies enable initial estimates of dose rates for small rodents, but no clinical trials have been carried out in these smaller laboratory species. Injectable (intravenous only) and oral preparations (tablets and powder) are available.

e. Flunixin. Flunixin has been reported as being effective in controlling post-operative pain in dogs (Reid and Nolan, 1991), and it has been widely used as an analgesic in larger species (cattle and horses). It also appears to be an effective analgesic in pigs, sheep and cats, but no controlled trial has been undertaken in these species. Both injectable and oral preparations are available. The most significant problem reported has been nephrotoxicity, either when flunixin was administered together with a known nephrotoxin agent (Mathews *et al.*, 1990), or in circumstances when renal blood flow was likely to have been compromised (McNeil, 1992). The mechanism of action has been suggested to be inhibition of the normal prostaglandin regulation of renal blood flow, resulting in a failure of renal perfusion during periods of hypotension. Good anaesthetic practice, appropriate fluid therapy and administration of flunixin after completion of surgery are likely to minimize this risk. Administration to conscious, healthy animals appears not to be associated with any significant risk.

f. Carprofen. Carprofen can provide effective post-surgical pain relief in the dog and the rat (Liles and Flecknell, 1993a; Nolan and Reid, 1993), and has also been used in a number of different species with apparent success. Both oral and injectable preparations are available.

g. Ketoprofen. Ketoprofen provides moderate pain relief in dogs, cats and horses. Its efficacy in other species is uncertain, although likely effective dose rates can be suggested from analgesiometric data. Both oral and injectable formulations are available.

h. Naproxen. Naproxen is unusual in having an exceptionally long half-life in the dog (35 hours) and has been used to alleviate moderate pain in this species, although, as with most analgesics, no controlled clinical trial has been undertaken. Naproxen is available as tablets and as an oral suspension.

i. Ketorolac. Ketorolac is used in humans to control moderate to severe post-operative pain, and early clinical trials suggest it may be effective in the dog (Mathews *et al.*, 1994). Both injectable and oral formulations are available. As with other NSAIDs, ketorolac is best not administered to animals with pre-existing renal disease or fluid deficits.

j. Meloxicam. Meloxicam is available in the UK as an oral suspension for use in dogs. It is likely to be effective against mild to moderate pain.

2. *Opioids (narcotic analgesics)*

A wide range of different opioid analgesics are available for use in animals. The different drugs vary in their analgesic potency, duration of action and also in their effects on other body systems. Opioids are classified by their activity at specific opioid receptors. The most clinically important of these are the mu and kappa receptors. Morphine is a mu agonist (it binds to and activates mu receptors). Mu antagonists (e.g. naloxone and naltrexone) bind to mu receptors but do not activate the receptor. Some analgesics are mu antagonists, and so will reverse the effects of mu agonists such as morphine, but also have agonist effects at kappa receptors. These agents are generally referred to as mixed agonist/antagonist analgesics (nalbuphine, butorphanol, pentazocine). Some opioid analgesics are classified as partial agonists that have agonist (analgesic) effects at the mu receptor, and also antagonize pure mu agonists (buprenorphine).

Opioid agonists and partial agonists relieve pain without impairing other sensations. However, they can cause some undesirable side-effects. All opioid agonists can produce some degree of respiratory depression, but when administered at clinically effective dose rates this is rarely a serious problem in animals. Opioids may also cause sedation or excitement, their effects varying considerably in different animal species. The effects on behaviour also depend upon the dose of drug that has been administered (Flecknell, 1984; Lumb and Wynn Jones, 1984).

When administered at dose rates appropriate for providing post-operative analgesia, opioids have minimal effect upon the cardiovascular system (Pircio *et al.*, 1976; Cowan *et al.*, 1977a; Popio *et al.*, 1978; Trim, 1983; O'Hair *et al.*, 1988). Higher dose rates, such as those that might be used when giving opioids as part of a balanced anaesthetic regimen, can cause bradycardia, although this can be prevented by administering atropine. In addition, morphine, pethidine and some other opioids can stimulate histamine release, which can produce a peripheral vasodilation. Clinically significant hypotension is usually seen only after administration of large doses or after rapid intravenous administration.

Opioids can cause vomiting in some animal species, notably in non-human primates and dogs. This side-effect is seen primarily when opioids are administered to pain-free animals (e.g. as pre-anaesthetic medication), and is less frequent when the drugs are administered post-operatively. In addition to causing vomiting, opioids may delay gastric emptying, decrease intestinal peristalsis and cause spasm of the biliary tract. These effects may preclude the use of opioids in certain experimental procedures, but generally the effects are of minimal clinical significance in animals. The detailed pharmacology of opioids has been extensively reviewed; general introductions to the field can be found in a number of sources (Bullingham, 1983; Smith and Covino, 1985).

a. Opioid agonists

i. Morphine. Morphine is obtained from opium and has been used as an analgesic in humans for many years. It has been extensively studied in a range of experimental animals and is also used in veterinary clinical practice (Hall and Clarke, 1991). Its duration of action in most animals is 2–4 hours, but a slow-release injectable preparation (Duromorph, Appendix 5) and a slow-release oral preparation (MST, Napp; Oramorph SR, Boehringer Ingelheim, Appendix 6) are also available. Rapid intravenous injection in the dog can cause transient hypotension because of histamine release, but this is not a problem if the drug is given by continuous intravenous infusion (see below). Although morphine remains one of the most useful and potent analgesics, it is relatively short-acting in many species (< 4 hours), and its administration after neuroleptanalgesic anaesthetic techniques in laboratory species can, not surprisingly, result in severe respiratory depression. It is also a drug with significant abuse potential.

ii. Pethidine (meperidine). Pethidine (meperidine) has been widely used as an analgesic in veterinary practice in the UK, but it has a relatively short duration of action in many species (< 2 hours). It has a spasmolytic action on smooth muscle in some species and this has led to its recommendation for use in specific clinical situations such as colic in horses. Both oral and injectable formulations are available.

iii. Methadone. Methadone has been used clinically as an analgesic in the horse, dog (Lumb and Wyn Jones, 1973; Hall and Clarke, 1991) and cat (Dobromylskyj, 1993), and dose rates for use in other species can be extrapolated from results of experimental analgesiometry (Flecknell, 1984). Both injectable and tablet formulations are available.

iv. Oxymorphone. Oxymorphone has actions similar to morphine and has been reported to be an effective analgesic in dogs and cats (Palminteri, 1963; Sawyer *et al.*, 1992).

v. Dextropropoxyphene. Dextropropoxyphene is a derivative of methadone which is used in humans for the relief of mild to moderate pain. In human clinical practice it is frequently administered in combination with aspirin or paracetamol and these preparations have occasionally been used in dogs in veterinary clinical practice in the UK (Yoxall, 1978).

vi. Codeine and dihydrocodeine. Codeine and dihydrocodeine are morphine derivatives of low and moderate potency, respectively. Codeine is used in combination with paracetamol for the relief of mild to moderate pain. Dihydrocodeine is also available as an oral preparation, and is an effective analgesic in humans. To date, no information concerning its clinical efficacy in animals is available.

vii. Fentanyl. Fentanyl is a potent, relatively short-acting synthetic opiate. Its main use in laboratory animal anaesthesia is in the neurolept-analgesic combinations Hypnorm (fentanyl/fluanisone) and Innovar Vet (fentanyl/droperidol). Because of its short duration of action — under 30 minutes in most species (Lumb and Wynn Jones, 1973) — fentanyl is most widely used for providing analgesia during surgical procedures (Andrews and Prys-Roberts, 1983). If it is to be used to control post-operative pain, it should be administered as a continuous infusion.

viii. Alfentanil. Alfentanil is a synthetic opioid related to fentanyl. It has similar pharmacodynamics to fentanyl, but has a more rapid onset and shorter duration of action. Alfentanil can be administered by continuous infusion to provide analgesia during surgical procedures, and its short duration of action enables good moment-to-moment control of the intensity of the analgesic effect.

b. Opioid mixed agonist/antagonists and partial agonists

i. Pentazocine. Pentazocine has been reported to provide effective analgesia in a range of animal species (Lumb and Wynn Jones, 1973; Cooper and Organ, 1977; Taylor and Houlton, 1984). It has been reported to produce dysphoria in humans (Rosow, 1985), but it is uncertain whether similar effects occur in animals. Generally, the sedative effect of pentazocine is less than that of morphine. Both oral and injectable formulations are available.

ii. Butorphanol. Butorphanol has a veterinary product licence as an analgesic in several countries, and is believed to provide post-operative analgesia in a variety of species (Sawyer and Rech, 1987; Flecknell and

Liles, 1990; Sawyer et al., 1991). Butorphanol has marked mu opioid antagonist properties and can be used to reverse the action of fentanyl while maintaining some analgesic effects by its action at kappa receptors.

iii. Buprenorphine. Buprenorphine is a potent partial mu agonist that has the advantage of having a prolonged duration of action in many species (Cowan *et al.*, 1977a,b; Dum and Herz, 1981; Nolan *et al.*, 1987; Flecknell and Liles, 1990), but the degree of analgesia produced may not always be sufficient in some individuals. The drug has been used in veterinary clinical practice in dogs and horses (Hall and Clarke, 1991), cats (Taylor, 1985) and in a wide range of laboratory animal species (Flecknell, 1991; Liles and Flecknell, 1993a). Buprenorphine is available as an injectable formulation, and as tablets for sublingual administration in humans.

iv. Nalbuphine. Nalbuphine has been reported to provide effective analgesia in dogs (Flecknell *et al.*, 1991) and rats (Flecknell and Liles, 1991). It has a duration of action of 2–4 hours in most species. Nalbuphine rapidly and effectively antagonizes the effects of mu agonists such as fentanyl, while maintaining an analgesic effect at kappa receptors. It is therefore particularly suitable for reversal of opioid-based anaesthetic regimens.

3. Clinical use of analgesics

A number of clinical problems arise when analgesics are administered to control post-operative pain. The most important problem is the short duration of action of most of the opioid (narcotic) analgesics. Maintenance of effective analgesia with, for example, pethidine, may require repeated administration every 1–3 hours, depending upon the species. Continuation of such a regimen overnight can cause practical problems. One method of avoiding this difficulty is to use buprenorphine as the analgesic, since there is good evidence in humans, rodents, rabbits and pigs that it has a duration of action of 6–12 hours (Cowan *et al.*, 1977a; Heel *et al.*, 1979; Dum and Herz, 1981; Hermansen *et al.*, 1986; Flecknell and Liles, 1990). In clinical use in a wide range of animal species, it appears to provide effective pain relief for 6–12 hours. Its duration of action in the sheep appears to be considerably less, although still of longer duration than pethidine and morphine (Nolan *et al.*, 1987).

An alternative approach is to adopt the well-established human clinical technique of administering analgesics as a continuous infusion. Infusions of analgesics have the advantage of maintaining effective plasma levels of the analgesic, thus providing continuous pain relief. This contrasts with intermittent injections, where pain may return before the next dose of analgesic is administered. This technique obviously poses some methodological difficulties in animals, but if an indwelling catheter and harness and swivel apparatus are available, then this can be arranged quite simply. In larger species (> 3–4 kg body weight), a lightweight infusion pump (Graseby Medical, Appendix 7) can be bandaged directly to the animal and continuous infusion made simply by means of a butterfly-type needle anchored subcutaneously or intramuscularly. When analgesics are to be administered by continuous infusion, the infusion rate can be calculated from a knowledge of the pharmacokinetics of the analgesic to be used (Mather, 1983) (see Chapter 5). If these data are not readily available, an approximation that appears successful in clinical use is as follows: calculate the total dose required over the period of infusion, reduce this by half and set the pump infusion rate accordingly; administer a single, normal dose of the drug as an initial loading dose and start the infusion. The rate can then be adjusted depending upon the animal's responses.

a. Oral administration. Repeated injections of analgesics are time-consuming and may be distressing to the animal, particularly smaller species that require firm physical restraint to enable an injection to be given safely and effectively. In addition, the need for repeated injections requires veterinary or other staff to attend the animal overnight. To circumvent this problem, the possibility of incorporating analgesics in food or water has been investigated (Kistler, 1988). Long-term analgesia can be produced by this route: Kistler (1988) reported that rats had demonstrable analgesia for a 2-week period when buprenorphine was administered continuously in the drinking water. Unfortunately, several practical problems limit the use of this technique. Some animals eat and drink relatively infrequently, or may only do so in the dark phase of their photoperiod. In addition, food and water intake may be depressed following surgery, and this, coupled with wide individual variation in consumption, makes routine application of the technique difficult. Finally, the high first-pass liver metabolism of opioids administered by the oral route requires that large doses are given, and this can represent a significant cost if all of the animals' drinking water or food is medicated.

Administration of small quantities of medicated food does not avoid the need for repeated attendance overnight, but does remove the need for repeated subcutaneous or intramuscular injections in small rodents. Provision of analgesia with buprenorphine in flavoured gelatin ('Buprenorphine Jell-O'; Pekow, 1992) seems to be an effective means of providing post-operative pain relief. In the author's laboratory, rats were found to be initially cautious of jelly pellets, but once one pellet had been consumed, subsequent pellets were eaten as soon as they were offered. It is therefore advisable to commence administering pellets that do not contain analgesic 2–3 days before surgery. After surgery, analgesic-containing jelly can be given. The flavoured gelatin used is domestic fruit-flavoured jelly, reconstituted at double the recommended strength. Rats receiving buprenorphine jelly following laparotomy had a significantly lower reduction in body weight, due to maintenance of a more normal pattern of food and water consumption, than untreated control animals (Liles, 1994). The dose rate used, 0.5 mg kg^{-1}, is approximately 10 times that given by subcutaneous injection, to compensate for first-pass metabolism in the liver after oral administration.

Techniques for administration of food pellets at intervals to experimental animals are well established, and it would be a relatively simple procedure to introduce an automated means of delivering pellets at appropriate time intervals. The technique could also be used with larger species, and need not be restricted to opioids, or indeed analgesics. Provided that the animal is eating or drinking, small quantities of highly palatable material could be provided at appropriate intervals. Simple timer devices to achieve this are already marketed for delayed feeding of pet dogs and cats.

b. Epidural and intrathecal opioids. Epidural and intrathecal opioids have been shown to have a prolonged effect in humans, and to provide effective analgesia (Glynn, 1987). In animals, both clinical and experimental studies have indicated that the technique can be used in a number of species (Dodman *et al.*, 1992; Duke *et al.*, 1993; Pablo, 1993; Pascoe, 1993; Popilskis *et al.*, 1993). Although used as a research tool in laboratory species (Yaksh *et al.*, 1988), this route of administration has yet to be exploited as a means of controlling post-operative pain. The necessary techniques of epidural or intrathecal injection have been described in the rabbit (Kero *et al.*, 1981; Hughes *et al.*, 1993) and guinea pig (Thomasson *et al.*, 1974). For larger species such as the cat, dog, sheep and pig, descriptions of the injection technique can be found in most veterinary anaesthesia text books and a number of other publications (Klide and Soma, 1968; Lumb and Wynn Jones, 1973; Hall and Clarke, 1991).

As mentioned above, the administration of opioids by any route can be associated with the development of respiratory depression. It must be emphasized that this is rarely of clinical significance in animals, unless high doses of pure mu agonists (e.g. fentanyl) are used. If respiratory depression occurs, it can be treated by the administration of the opiate antagonist drug, naloxone. Administration of naloxone will also reverse the analgesic effects of the opioid and it may be preferable to correct the respiratory depression by the use of doxapram. Alternatively, if a mu agonist opioid such as morphine or fentanyl has been used, the respiratory depression can be reversed using nalbuphine or butorphanol, and some analgesia maintained because of the action of these last two agents at kappa receptors. Repeated administration of these agents may be required, and the animal should be observed carefully for several hours to ensure adequate respiratory function is maintained.

4. Additional considerations in pain relief

Although the use of analgesic drugs remains the most important technique for reducing post-operative pain, the use of these drugs must be integrated into a total scheme for peri-operative care. As discussed in Chapter 1, pain relief in the immediate recovery period can be provided by including an analgesic drug in any pre-anaesthetic medication. Alternatively, if a neuroleptanalgesic combination has been used to produce anaesthesia, it can be reversed by the use of buprenorphine, nalbuphine or butorphanol, rather than with naloxone. These agents have been shown not only to reverse the respiratory depressant effects of opioids such as fentanyl, but also (in contrast to naloxone) to provide effective prolonged analgesia (Robertson and Laing, 1980; Latasch *et al.*, 1984; Flecknell *et al.*, 1989a).

The expertise of the surgeon can also greatly influence the degree of post-operative pain. Good surgical technique, which minimizes tissue trauma, and the prevention of tension on suture lines can considerably reduce post-operative pain. The use of bandages to pad and protect traumatized tissue must not be overlooked and forms an essential adjunct to the use of analgesic drugs.

Along with measures directed towards alleviating or preventing pain, it is important to consider the overall care of the animal and the prevention of distress. 'Distress' is used in this context to describe conditions which are not in themselves painful, but which are unpleasant and which the animal would normally choose to avoid. For example, recovering from anaesthesia on wet, uncomfortable bedding in a cold, unfamiliar environment would be likely to cause distress to many animals. It is essential to consider the methods described for the control of pain in conjunction with the techniques discussed earlier aimed at providing good post-operative care.

5. Recommendations

It is difficult to make firm recommendations concerning which analgesics to use routinely, and how often to give them, because of the various factors outlined above. Nevertheless, as a general guide, the following techniques are used routinely in the author's research facility.

When carrying out any surgical procedure, buprenorphine is administered either pre-operatively or immediately following induction of anaesthesia, if a volatile anaesthetic is used. If neuroleptanalgesic regimens are used, or mu opioids are given as part of a balanced anaesthetic technique, then administration of buprenorphine is delayed until completion of surgery. If the procedure is relatively minor, for example jugular or carotid cannulation, then only a single dose of analgesic is administered. In some circumstances a potent NSAID, such as flunixin or carprofen, may be used as an alternative to buprenorphine.

Following more invasive surgical procedures, such as laparatomy, orthopaedic surgery or craniotomy, opioid administration is continued for 24–48 hours. When major surgery is undertaken, particularly in larger species when the degree of tissue trauma tends to be greater, analgesic administration may continue for 72 hours. Frequently, the regimen chosen consists of opioids (buprenorphine) in combination with an NSAID for 24–36 hours, followed by NSAID alone for a further 24–36 hours (see Table 6.3 for suggested dose rates).

IV. CONCLUSIONS

Attention to the suggestions made in this chapter concerning post-operative care can have a dramatic effect on the speed with which animals return to normality following surgical procedures. It has been repeatedly demonstrated in humans that the provision of effective analgesia reduces the time taken for post-operative recovery (Smith and Covino, 1985). The provision of good post-operative care should be considered essential both because of a concern for the animal's welfare and also because it is good scientific practice.

7

Anaesthesia of common laboratory species

Laboratory animals are anaesthetized either to provide humane restraint while relatively atraumatic procedures are carried out, or to eliminate the perception of pain during surgical operations. When selecting a method of anaesthesia, concern for the welfare of the animal requires that administration of the anaesthetic should cause a minimum of distress. In addition, the drugs used should provide effective analgesia and an uneventful recovery, free from unpleasant side-effects. These considerations must often be balanced against the research worker's requirements for ease of administration of the anaesthetic, good recovery rate and the use of drugs that have a minimum influence on the experimental work that is being undertaken.

These factors have been considered carefully in selecting the methods recommended for each species in this chapter. The primary consideration has been the well-being of the animal, coupled with ease of use and safety of the drug or drug combination. For a more extensive discussion of selecting an anaesthetic agent, see Chapter 3.

Alternative anaesthetic regimens are included since some research protocols will preclude the use of the drugs recommended, and, in addition, some drugs may not be readily available in certain laboratories. A comprehensive listing of anaesthetic drug dose rates for each species is provided.

It is particularly important to read the notes below on individual species in conjunction with the general chapters on intra-operative care.

The recommended dose rates are those that have been found effective in the majority of individuals of the species concerned. The response to an anaesthetic drug can vary considerably and may be influenced by the strain of animal, sex, age and environmental conditions in which the animal is housed (Green, 1981b; Lovell, 1986a–c). When using a drug or drug combination for the first time, or when anaesthetizing a different strain of animal, it is advisable to proceed cautiously. As experience is gained, a dose rate appropriate to the particular strain can be established.

In order to provide some guidance as to the predicted effect of the different anaesthetics and their duration of action, the dose rate tables for each species include an estimate of depth of anaesthesia, its duration, and the likely duration of loss of the animal's righting reflex. It is important to note that considerable variation in response is to be expected. With many agents, a range of dose rates is given with a corresponding range of anticipated effects (e.g. light or deep anaesthesia). The terminology used is as follows:

Sedation (light, medium or heavy): the animal will have reduced activity and may become completely immobile, but is easily aroused, particularly by painful stimuli.

Analgesia: some pain-alleviating effect is present.

Immobilization: the animal is immobilized but still responds to painful stimuli.

Light anaesthesia: the animal is immobile and unconscious, but still responsive to even minor surgical procedures.

Medium anaesthesia: most surgical procedures (e.g. laparotomy) may be carried out without causing any response, but the animal may still respond to major surgical stimuli (e.g. orthopaedic surgery).

Deep anaesthesia: the animal is unresponsive to all surgical stimuli.

These terms are used to provide a general guide, but in all instances the depth of anaesthesia produced in a particular animal should be assessed before commencing surgery (see Chapter 4).

I. SMALL RODENTS

The problems that arise when anaesthetizing rodents are related primarily to the small body size of these species. Their high ratio of surface area to body weight makes them particularly susceptible to the development of hypothermia; intravenous drug administration is limited by the size of the superficial veins and the small and relatively inaccessible larynx makes endotracheal intubation difficult. A further consequence of the small size of these species is that the volumes of anaesthetic required may be very small. In many instances it will be found convenient to mix together the required compounds and dilute them with sterile water for injection. Suggested dilutions are given in Appendix 3.

An additional complication is the occurrence of subclinical lung disease in many laboratory colonies of rats and mice. This may cause respiratory failure during the period of anaesthesia or result in the development of severe clinical respiratory disease in the post-operative period. Nevertheless, it is possible to achieve safe and effective anaesthesia in these species, provided that special attention is given to their particular requirements.

It is unnecessary to withhold food and water before induction of anaesthesia since vomiting on induction or recovery does not occur in any of the small rodents. As mentioned in Chapter 1, problems may be seen with some guinea pigs that retain food in their pharynx after being anaesthetized. If this occurs then a short period of pre-anaesthetic fasting (6–8 hours) should be introduced.

A. Rats

1. Pre-anaesthetic medication

Provided that the rat has become accustomed to being handled, most animals can easily be restrained humanely to enable the intraperitoneal or intramuscular injection of an anaesthetic agent. In general, intraperitoneal injections are tolerated better than intramuscular injections as they cause less pain to the animal. Pre-anaesthetic medication to sedate the animal is therefore not usually required but, if an intravenous induction agent is to be used, initial sedation with a tranquillizer or sedative/analgesic is recommended.

The following drugs can be used to produce sedation and are listed in order of preference (Table 7.1).

1. Fentanyl/fluanisone (Hypnorm, Janssen) (0.2–0.5 ml kg^{-1} i/m; 0.3–0.6 ml kg^{-1} i/p). At the lower dose rate sedation and some analgesia are produced. The higher dose rate produces sufficient analgesia to enable procedures such as skin biopsy or cardiac puncture to be carried out (Green, 1975). Occasionally, marked respiratory depression is seen when the drug is administered at the higher dose rate. If this produces severe cyanosis, it can be reversed with nalbuphine, butorphanol or naloxone.

2. Medetomidine (30–100 μg kg^{-1} s/c) produces light to heavy sedation and at the higher dose rate many animals will lose their righting reflex. Some strains require significantly higher dose rates (300 μg kg^{-1}) before becoming sedated and losing their righting reflex. The degree of analgesia produced is insufficient for anything other than very minor procedures, but is suitable for non-painful manipulations such as radiography (Virtanen, 1989). Medetomidine markedly potentiates the effects of other anaesthetic agents. For example, the concentration of volatile agent needed to produce surgical anaesthesia may be reduced by more than 60%.

TABLE 7.1
Sedatives, tranquillizers and other pre-anaesthetic medication for use in the rat

Drug	Dose rate	Comments
Acepromazine	2·5 mg kg^{-1} i/m, i/p	Light sedation
Atropine	0·05 mg kg^{-1} i/p, s/c	Anticholinergic
Diazepam	2·5–5·0 mg kg^{-1} i/p, i/m	Light sedation
Fentanyl/dropiderol (Innovar Vet)	0·5 ml kg^{-1} i/m	Immobilization/analgesia
Fentanyl/fluanisone (Hypnorm)	0·2–0·5 ml kg^{-1} i/m 0·3–0·6 ml kg^{-1} i/p	Light/moderate sedation, moderate analgesia
Glycopyrrolate	0·5 mg kg^{-1} i/m	Anticholinergic
Ketamine	50–100 mg kg^{-1} i/m, i/p	Deep sedation, immobilization, mild analgesia
Medetomidine	30–100 μg kg^{-1} s/c, i/p	Light to heavy sedation, some analgesia
Midazolam	5 mg kg^{-1} i/p	Light sedation
Xylazine	1–5 mg kg^{-1} i/m, i/p	Light to heavy sedation, some analgesia

Considerable variation in effect occurs between different strains.

3. Xylazine (1–5 mg kg^{-1} i/m or i/p) produces light to moderate sedation. Although the drug provides little analgesia when used alone, it markedly potentiates the effects of other anaesthetic agents.
4. Ketamine (50–100 mg kg^{-1} i/m or i/p) produces deep sedation. The degree of muscle relaxation is poor and the level of analgesia is insufficient for even superficial surgery (Green *et al.*, 1981a).
5. Acepromazine (2.5 mg kg^{-1} i/m or i/p) produces light sedation, but has no analgesic action.
6. Diazepam or midazolam (2.5–5 mg kg^{-1} i/m or i/p) produces light sedation, but neither drug has any analgesic action.

Atropine (0.05 mg kg^{-1} i/p or s/c) or glycopyrrolate (0.5 mg kg^{-1} i/m) (Olson *et al.*, 1993) can be administered to reduce salivary and bronchial secretions and protect the heart from vagal inhibition.

Appropriate pre-anaesthetic medication will reduce the stress caused by induction of anaesthesia and also ease handling and restraint. In addition, it will usually reduce the amount of other anaesthetic agents required to produce general anaesthesia. The dose rates of anaesthetic drugs quoted in Table 7.2 apply to rats which have received no pre-anaesthetic medication unless otherwise stated. Generally, these dosages can be reduced by at least 30–50% if one of the drugs listed above has been administered.

2. General anaesthesia

a. Injectable agents. The small body size of the rat makes intravenous injection difficult and drugs are usually administered by the intraperitoneal or intramuscular routes. If these routes are used, it is not possible to administer the drug gradually to effect and the anaesthetic must be given as a single, calculated dose. Because of the wide variation in drug response between different strains of rat, between male and female animals and between individuals, it is best to use a drug or drug combination that provides a wide margin of safety. Anaesthetic dose rates are summarized in Table 7.2.

1. The anaesthetic combination of choice for rats is fentanyl/fluanisone (Hypnorm, Janssen) together with diazepam or midazolam (0.6 ml kg^{-1} i/p Hypnorm, and diazepam 2.5 mg kg^{-1} i/p). When using midazolam the components are mixed together with water for injection (see Appendix 3). These combinations provide good surgical anaesthesia with excellent muscle relaxation lasting about 20–40 minutes (Green, 1975; Flecknell and Mitchell, 1984). Longer periods of anaesthesia can be achieved by the administration of additional doses of Hypnorm (about 0.1 ml kg^{-1} i/m every 30–40 min). Following the completion of surgery, the anaesthesia can be reversed using nalbuphine (0.1 mg kg^{-1} i/v, 1.0 mg kg^{-1} i/p or s/c) or butorphanol (0.1 mg kg^{-1} i/v, 2 mg kg^{-1} i/p or s/c).

2. A second safe and effective anaesthetic regimen is fentanyl (300 μg kg^{-1} i/p) and medetomidine (300 μg kg^{-1} i/p). The two agents can be mixed and administered as a single injection. Fentanyl/medetomidine provides about 60 minutes of surgical anaesthesia (Hu *et al.*, 1992). Recovery is very rapid provided that anaesthesia is reversed by administration of atipamezole (1 mg kg^{-1} s/c or i/p) (to reverse the medetomidine) and either nalbuphine (0.1 mg kg^{-1} i/v, 1.0 mg kg^{-1} i/p or s/c), butorphanol (0.1 mg kg^{-1} i/v, 2 mg kg^{-1} i/p or s/c) or another mixed agonist/antagonist opioid analgesic (see Tables 7.3 and 6.3). Experience has shown that the quality of induction and recovery with this method of anaesthesia is greatly improved by allowing the rats to acclimatize for 1–2 hours after movement to the room in which the procedure will be undertaken (Drage J. 1995, personal communication).

3. A third effective alternative is to administer medetomidine (0.5 mg kg^{-1} i/p) or xylazine (10 mg kg^{-1} i/p) and ketamine (75 mg kg^{-1} i/p). The two compounds can be mixed in the same syringe, and provide good

TABLE 7.2
Anaesthetic dose rates in the rat

Drug	Dose rate	Effect	Duration of anaesthesia (min)	Sleep time (min)
Alphaxalone/alphadolone	10–12 mg kg^{-1} i/v	Surgical anaesthesia	5	10
Chloral hydrate	400 mg kg^{-1} i/p	Light/surgical anaesthesia	60–120	120–180
Alpha-chloralose	55–65 mg kg^{-1} i/p	Light anaesthesia	480–600	Non-recovery only
Etorphine/methotrimeprazine (Immobilon) + midazolam	0·5 ml kg^{-1}s/c*	Surgical anaesthesia	60–70	120–240
Fentanyl/fluanisone + diazepam	0·6 ml kg^{-1} i/p + 2·5 mg kg^{-1} i/p	Surgical anaesthesia	20–40	120–240
Fentanyl/fluanisone/midazolam	2·7 ml kg $^{-1}$ i/p†	Surgical anaesthesia	20–40	120–240
Fentanyl + medetomidine	300 μg $^{-1}$ kg + 300 μg kg^{-1} i/p	Surgical anaesthesia	60–70	240–360
Inactin	80 mg kg^{-1} i/p	Surgical anaesthesia	60–240	120–300
Ketamine/acepromazine	75 mg kg^{-1} + 2·5 mg kg^{-1} i/p	Light anaesthesia	20–30	120
Ketamine/diazepam	75 mg kg^{-1} + 5 mg kg^{-1} i/p	Light anaesthesia	20–30	120
Ketamine/medetomidine	75 mg kg^{-1} + 0·5 mg kg^{-1} i/p	Surgical anaesthesia	20–30	120–240

Drug	Dose rate	Effect	Duration of anaesthesia (min)	Sleep time (min)
Ketamine/midazolam	75 mg kg^{-1} + 5 mg kg^{-1} i/p	Light anaesthesia	20–30	120
Ketamine/xylazine	75–100 mg kg^{-1} + 10 mg kg^{-1} i/p	Surgical anaesthesia	20–30	120–240
Methohexitone	10–15 mg kg^{-1} i/v	Surgical anaesthesia	5	10
Pentobarbitone	40–50 mg kg^{-1} i/p	Light anaesthesia	15–60	120–240
Propofol	10 mg kg^{-1} i/v	Surgical anaesthesia	5	10
Thiopentone	30 mg kg^{-1} i/v	Surgical anaesthesia	10	15
Tiletamine/zolezepam	40 mg kg^{-1} i/p	Light anaesthesia	15–25	60–120
Urethane	1000 mg kg^{-1} i/p	Surgical anaesthesia	360–480	Non-recovery only

* Dose in ml kg^{-1} of a mixture of 1 part Immobilon, 1 part midazolam (5 mg ml^{-1} initial concentration) and 2 parts water for injection.

† Dose in ml kg^{-1} of a mixture of 1 part Hypnorm plus 2 parts water for injection, and 1 part midazolam (5 mg ml^{-1} initial concentration).

Duration of anaesthesia and sleep time (loss of righting reflex) are provided only as a general guide, since considerable between animal variation occurs. For recommended techniques, see text.

TABLE 7.3
Antagonists to anaesthetic regimens for use in rodents and rabbits

Compound	Anaesthetic regimen	Dose rate	Comments
Atipamezole	Any regimen using xylazine or medetomidine	0·1–1 mg kg^{-1} i/m, i/p, s/c, or i/v	Highly specific antagonist. Dose required varies depending on dose of xylazine or medetomidine administered
Buprenorphine	Any regimen using mu opioids	See Table 6.3c	Slower onset than naloxone and nalbuphine, but longer-acting
Doxapram	All anaesthetics	5–10 mg kg^{-1} i/m, i/v or i/p	General respiratory stimulant
Flumazenil	Benzodiazepine (e.g. midazolam)		Dose varies depending upon dose of benzodiazepine. Resedation may occur
Nalbuphine	Any regimen using mu opioids	See Table 6.3c	Almost as rapid-acting as naloxone, maintains post-operative analgesia
Naloxone	Any regimen using mu opioids (e.g. fentanyl)	0·01–0·1 mg kg^{-1} i/v, i/m or i/p	Reverses analgesia and respiratory depression
Yohimbine	Any regimen using xylazine or medetomidine	0·2 mg kg^{-1} i/v, 0·5 mg kg^{-1} i/m	Relatively non-specific antagonist. Not recommended

surgical anaesthesia, although the depth of anaesthesia may be insufficient for major surgery in some animals (Van-Pelt, 1977; Green *et al.*, 1981a; Hsu *et al.*, 1986; Wixson *et al.*, 1987; Nevalainen *et al.*, 1989). This combination provides about 30 minutes of surgical anaesthesia. The combination can be partially reversed using atipamezole (1 mg kg^{-1} s/c or i/p), but early reversal (10–20 minutes after induction) may be associated with undesirable behavioural disturbances owing to the effects of the ketamine (Morris T. H., 1995 personal communication).

4. If intravenous administration of drugs is feasible, then propofol (10 mg kg^{-1} i/v) (Glen, 1980; Brammer *et al.*, 1992; Cockshott *et al.*, 1992) or alphaxalone/alphadolone (10–12 mg kg^{-1} i/v) (Green *et al.*, 1978) produces surgical anaesthesia; both compounds are especially useful for administering by continuous infusion to provide stable, long-lasting anaesthesia. When administered by the intraperitoneal route the effects are less predictable and these drugs can only be relied upon to produce heavy sedation.

Alternative agents for intravenous induction are thiopentone (30 mg

kg^{-1} i/v, 1.25% solution) or methohexitone (10–15 mg kg^{-1} i/v, 1% solution) which can be used to produce 5–10 minutes of anaesthesia.

5. Pentobarbitone should be diluted to provide a 30 mg ml^{-1} solution and up to 40–50 mg kg^{-1} administered i/p. Severe respiratory depression invariably accompanies the onset of surgical anaesthesia and because of the consequent high mortality rate this drug is best avoided in rats.

b. Inhalational agents. The most convenient method of inducing anaesthesia in the rat is to use an anaesthetic chamber. This should be constructed from Perspex, so that the animal can be observed during induction. Anaesthetic vapour should be supplied from an anaesthetic machine and the chamber should be designed so that excess anaesthetic gas can be ducted to a gas-scavenging device or removed from the room via the ventilation system. Following induction of anaesthesia, the rat should be removed from the chamber and anaesthesia maintained using a small face-mask connected to the anaesthetic machine (see Chapter 3).

c. Endotracheal intubation. The major disadvantage of using a face-mask for connection of the animal to the anaesthetic gas supply is that it is difficult to assist ventilation should this prove necessary. Endotracheal intubation, together with use of an appropriate anaesthetic circuit, allows easy control of ventilation. Intubation is a difficult procedure in the rat, requiring special equipment. Further details of the technique are given in Chapter 3.

d. Mechanical ventilation. Several different ventilators are available for use in rats, with a variety of mechanisms of action. All aim to achieve controlled ventilation of the lungs by means of the application of intermittent positive pressure to the patient's airway. The ventilators manufactured by Harvard Apparatus Ltd (Appendix 7) are the most widely used in the UK. A volume-cycled piston model and a pressure-cycled ventilator are available (see Table 5.2). Neither of these ventilators has facilities for humidification of anaesthetic gases, but this can be achieved readily by bubbling the gases through a water bath. An economical approach is to purchase a human infant feeding-bottle warmer, fill it with water and pipe anaesthetic gases through a glass aerator placed in the unit.

3. Anaesthetic management

It is particularly important to prevent the development of hypothermia in rats. Ideally, the rat should be placed on a thermostatically controlled heating blanket (Harvard Apparatus Ltd); alternatively, heating lamps can be used. Both measures can be combined with the use of insulating material such as cotton wool, aluminium foil or bubble packing (Chapter 4). These measures must be continued in the post-operative recovery period.

B. Mice

1. Pre-anaesthetic medication

Mice are easily restrained humanely and it will rarely be necessary to produce sedation before induction of anaesthesia. If sedation is required, the following drugs can be used (Table 7.4):

1. Fentanyl/fluanisone (Hypnorm, Janssen) (0.1–0.3 ml kg^{-1} i/p) provides sedation and sufficient analgesia for superficial procedures such as ear punching (Green, 1975). The drug is most conveniently administered as a 1:10 dilution of the commercial preparation. The effects of this mixture can be partially reversed using nalbuphine (4 mg kg^{-1} s/c or i/p) or butorphanol (2 mg kg^{-1} i/p or s/c).
2. Medetomidine (30–100 µg kg^{-1} i/p) produces light to deep sedation.

TABLE 7.4
Sedatives, tranquillizers and other pre-anaesthetic medication for use in the mouse

Drug	Dose rate	Comments
Acepromazine	2–5 mg kg^{-1} i/p, s/c	Light sedation
Atropine	0·04 mg kg^{-1} s/c	Anticholinergic
Diazepam	5 mg kg^{-1} i/m, i/p	Light sedation
Fentanyl/droperidol (Innovar Vet)	0·5 ml kg^{-1} i/m	Immobilization, analgesia
Fentanyl/fluanisone (Hypnorm)	0·1–0·3 ml kg^{-1} i/p	Light sedation, moderate analgesia
Ketamine	100–200 mg kg^{-1} i/m	Deep sedation, mild analgesia
Medetomidine	30–100 µg kg^{-1} s/c	Light to deep sedation, some analgesia
Midazolam	5 mg kg^{-1} i/m, i/p	Light to moderate sedation
Xylazine	5–10 mg kg^{-1} i/p	Light sedation, some analgesia

Considerable variation in effects occurs between different strains.

As with the rat, considerable strain variation may occur. Sedation can be completely reversed using atipamezole (1 mg kg^{-1} i/p).

3. Xylazine (5–10 mg kg^{-1} i/p) produces sedation but appears to have little analgesic action when used alone in mice.

4. Acepromazine (2–5 mg kg^{-1} i/p) produces sedation but has no analgesic action.

5. Diazepam (5 mg kg^{-1} i/p) or midazolam (5 mg kg^{-1} i/p) produces sedation but no analgesia.

Atropine (0.04 mg kg^{-1} s/c, i/m or i/p) can be administered to reduce salivary gland and bronchial secretions. Dose rates for general anaesthetics given below should be reduced by 30–50% if one of the sedative drugs listed above has been administered.

2. *General anaesthesia*

a. Injectable agents. As with the rat, drugs are most conveniently administered by the intraperitoneal route. Dose rates of anaesthetic agents are summarized in Table 7.5.

1. The anaesthetic combination of choice is fentanyl/fluanisone (Hypnorm, Janssen) together with midazolam or diazepam (0.4 ml kg^{-1} i/p Hypnorm, and diazepam 5 mg kg^{-1} i/p). When using midazolam the components are mixed together with water for injection (see Appendix 3). These combinations provide good surgical anaesthesia lasting about 30–40 minutes (Green, 1975; Flecknell and Mitchell, 1984). Anaesthesia can be prolonged by the administration of additional doses of Hypnorm (0.3 ml kg^{-1} every 30–40 min). Following the completion of surgery, anaesthesia can be partially reversed by the administration of nalbuphine (4 mg kg^{-1} i/p or s/c) or butorphanol (2.0 mg kg^{-1} i/p or s/c).

2. Ketamine (75 mg kg^{-1} i/p) and medetomidine (1.0 mg kg^{-1} i/p) produce moderate surgical anaesthesia in most strains of mouse. In some strains, however, the degree of analgesia is insufficient for major surgery such as laparotomy (Voipio *et al.*, 1990). The two compounds can be mixed in the same syringe and given as a single injection. Anaesthesia can be partially reversed by administration of atipamezole (1 mg kg^{-1} s/c).

3. An alternative combination is ketamine (80–100 mg kg^{-1} i/p) and xylazine (10 mg kg^{-1} i/p). This combination can be pre-mixed in the correct proportions and administered as a single i/p injection. It provides 20–30 minutes of anaesthesia, but the depth of anaesthesia is often insufficient to enable major surgery to be carried out humanely (Mulder

TABLE 7.5
Anaesthetic dose rates in the mouse

Drug	Dose rate	Effect	Duration of anaesthesia (min)	Sleep time (min)
Alphaxalone/alphadolone	10–15 mg kg^{-1} i/v	Surgical anaesthesia	5	10
Fentanyl/fluanisone (Hypnorm) + diazepam	0·4 ml kg^{-1} i/p + 5 mg kg^{-1} i/p	Surgical anaesthesia	30–40	120–240
Fentanyl/fluanisone (Hypnorm)/midazolam	10·0 ml kg^{-1} i/p*	Surgical anaesthesia	30–40	120–240
Ketamine/acepromazine	100 mg kg^{-1} + 5 mg kg^{-1} i/p	Immobilization/anaesthesia	20–30	40–120
Ketamine/diazepam	100 mg kg^{-1} + 5 mg kg^{-1} i/p	Immobilization/anaesthesia	20–30	60–120
Ketamine/medetomidine	75 mg kg^{-1} + 1·0 mg kg^{-1} i/p	Surgical anaesthesia	20–30	60–120
Ketamine/midazolam	100 mg kg^{-1} + 5 mg kg^{-1} i/p	Immobilization/anaesthesia	20–30	60–120
Ketamine/xylazine	80–100 mg kg^{-1} + 10 mg kg^{-1} i/p	Surgical anaesthesia	20–30	60–120
Methohexitone	10 mg kg^{-1} i/v	Surgical anaesthesia	5	10
Metomidate/fentanyl	60 mg kg^{-1} + 0·06 mg kg^{-1} s/c	Surgical anaesthesia	40–60	90–120
Pentobarbitone	40–50 mg kg^{-1} i/p	Immobilization/anaesthesia	20–40	120–180
Propofol	26 mg kg^{-1} i/v	Surgical anaesthesia	5–10	10–15
Thiopentone	30–40 mg kg^{-1} i/v	Surgical anaesthesia	5–10	10–15
Tiletamine/zolezepam	80 mg kg^{-1} i/p	Immobilization		60–120
Tribromoethanol	240 mg kg^{-1} i/p	Surgical anaesthesia	15–45	60–120

* Dose in ml kg^{-1} of a mixture of 1 part 'Hypnorm' plus 2 parts water for injection, and 1 part midazolam (5 mg ml^{-1} initial concentration).

Duration of anaesthesia and sleep time (loss of righting reflex) are provided only as a general guide, since considerable between animal variation occurs. For recommended techniques, see text.

and Mulder, 1979; Green *et al.*, 1981a; Erhardt *et al.*, 1984). Anaesthesia can be partially reversed by administration of atipamezole (1 mg kg^{-1} s/c).

4. The combination of metomidate (60 mg kg^{-1}) and fentanyl (0.06 mg kg^{-1}) produces stable surgical anaesthesia in mice (Green *et al.*, 1981b). The two drugs are combined and given as a single subcutaneous injection.

5. If the technique of intravenous injection can be mastered, then either propofol (26 mg kg^{-1} i/v) (Glen, 1980) or alphaxalone/alphadolone (10–15 mg kg^{-1} i/v) (Child *et al.*, 1971; Green *et al.*, 1978) can be used to provide short periods (5–10 minutes) of anaesthesia. An advantage of these compounds is that repeated administration to prolong anaesthesia is not associated with prolongation of the recovery period (see section on long-term anaesthesia in Chapter 5). Thiopentone (30–40 mg kg^{-1} i/v) or methohexitone (10 mg kg^{-1} iv)can also be used to provide short (5 minutes) periods of anaesthesia.

6. If pentobarbitone is to be used, it should be diluted to provide a 6 mg ml^{-1} solution and administered at a dosage of 40–50 mg kg^{-1} i/p. The variation of effect in different strains of mice is considerable, sleep times ranging from 50 minutes to 250 minutes (Lovell, 1986b) with identical doses of anaesthetic (60 mg kg^{-1}), so that overdosage or underdosage with this drug frequently occurs.

b. Inhalational agents. Induction of anaesthesia using an anaesthetic chamber is simple and convenient. Maintenance using a face-mask is straightforward, but may require a suitably sized mask to be constructed, for example from the end of a disposable syringe barrel.

c. Endotracheal intubation. Endotracheal intubation is difficult in mice and requires the use of a purpose-built laryngoscope (Costa *et al.*, 1986). Alternatively, an otoscope technique can be used, as in the rat (DeLeonardis *et al.*, 1995)

d. Mechanical ventilation. Mechanical ventilation can be performed using a Harvard Rodent Ventilator (Appendix 7) — see Chapter 5.

3. Anaesthetic management

Mice are even more prone than rats to develop hypothermia, and it is essential to take measures to maintain body temperature (see Chapter 4).

C. Hamsters

1. Pre-anaesthetic medication

Hamsters are easily restrained humanely and pre-anaesthetic sedation is rarely necessary. If restraint is a problem, an anaesthetic chamber should be used for induction of anaesthesia. If sedation is required, the following drugs can be used (Table 7.6).

1. Fentanyl/fluanisone (Hypnorm, Janssen) (0.5 ml kg^{-1} i/p) provides sufficient analgesia for superficial procedures.
2. Medetomidine (100 µg kg^{-1} s/c) produces moderate sedation in hamsters, but animals do not lose their righting reflex even at high dose rates (Morris, 1991).
3. Diazepam (5 mg kg^{-1} i/p) or midazolam (5 mg kg^{-1} i/p) produces sedation but no analgesia.

Atropine (0.04 mg kg^{-1} s/c, i/m or i/p) can be administered to reduce salivary and bronchial secretions. Dose rates for general anaesthesia given below should be reduced by 30–50% if one of the sedative drugs listed above has been administered.

TABLE 7.6
Sedatives, tranquillizers and other pre-anaesthetic medication for use in the hamster

Drug	Dose rate	Comments
Acepromazine	5 mg kg^{-1} i/p	Light sedation
Atropine	0·04 mg kg^{-1} s/c	Anticholinergic
Diazepam	5 mg kg^{-1} i/m, i/p	Light to moderate sedation
Fentanyl/droperidol (Innovar Vet)	0·9 ml kg^{-1} i/m	analgesia, unpredictable degree of sedation
Fentanyl/fluanisone (Hypnorm)	0·5 ml kg^{-1} i/p	Light to moderate sedation, moderate analgesia
Ketamine	100–200 mg kg^{-1} i/m	Deep sedation, mild analgesia
Medetomidine	100 µg kg^{-1} s/c or i/p	Moderate sedation, some analgesia
Midazolam	5 mg kg^{-1} i/m, i/p	Light to moderate sedation
Xylazine	5 mg kg^{-1} i/m	Light sedation, some analgesia

Considerable variation in effect occurs between different strains.

2. *General anaesthesia*

a. Injectable agents. As with other small rodents, drugs are most conveniently administered by intraperitoneal injection (Table 7.7).

1. The anaesthetic combination of choice is fentanyl/fluanisone (Hypnorm, Janssen) with midazolam or diazepam (1.0 ml kg^{-1} i/p Hypnorm, and diazepam 5 mg kg^{-1} i/p). When using midazolam the components are mixed together with water for injection (see Appendix 3). These combinations provide good surgical anaesthesia lasting about 20–40 minutes (Flecknell and Mitchell, 1984) and can be partially reversed with nalbuphine (2 mg kg^{-1} s/c) or butorphanol (2.0 mg kg^{-1} s/c).
2. An alternative, equally satisfactory, combination in the hamster is ketamine (200 mg kg^{-1} i/p) and xylazine (10 mg kg^{-1} i/p), which in this species appears to produce surgical anaesthesia reliably (Curl and Peters, 1983). Ketamine 100 mg kg i/p and medetomidine 250 mg kg i/p also appear to produce effective surgical anaesthesia (Townsend 1996). Anaesthesia can be partially reversed using atipamezole (1 mg kg s/c).
3. The use of pentobarbitone (50–90 mg kg^{-1} i/p) in hamsters is particularly hazardous and death often occurs. If pentobarbitone is to be used in any small rodent, it is best to administer a dose sufficient to produce light anaesthesia (50 mg kg^{-1} i/p) and then administer a volatile anaesthetic to produce full surgical anaesthesia.

b. Inhalational agents. Induction of anaesthesia using an anaesthetic chamber is simple and convenient, and halothane, isoflurane and methoxyflurane all provide effective and safe anaesthesia. A suitably sized mask may need to be constructed for anaesthetic maintenance, or a commercially available system purchased (e.g. International Market Supply, Appendix 7).

c. Endotracheal intubation. Endotracheal intubation is difficult in hamsters and requires the use of a purpose-built laryngoscope (Costa *et al.*, 1986).

d. Mechanical ventilation. Mechanical ventilation can be performed using a Harvard Rodent Ventilator (Appendix 7) — see Chapter 5.

3. *Anaesthetic management*

As with other small rodents, prevention of hypothermia is of critical importance (see Chapter 4).

TABLE 7.7
Anaesthetic dose rates in the hamster

Drug	Dose rate	Effect	Duration of anaesthesia (min)	Sleep time (min)
Alpha-chloralose	80–100 mg kg^{-1} i/p	Immobilization		180–240
Alphaxalone/alphadolone	150 mg kg^{-1} i/p	Immobilization/anaesthesia	20–60	120–150
Fentanyl/fluanisone (Hypnorm) + diazepam	1 ml kg^{-1} i/m or i/p + 5 mg kg^{-1} i/p	Surgical anaesthesia	20–40	60–90
Fentanyl/fluanisone (Hypnorm)/midazolam	4·0 ml kg^{-1} i/p*	Surgical anaesthesia	20–40	60–90
Ketamine/acepromazine	150 mg kg^{-1} + 5 mg kg^{-1} i/p	Immobilization/anaesthesia	45–120	75–180
Ketamine/diazepam	70 mg kg^{-1} + 2 mg kg^{-1} i/p	Immobilization/anaesthesia	30–45	90–120
Ketamine/medetomidine	100 mg kg^{-1} + 250 µg kg^{-1} i/p	Surgical anaesthesia	30–60	60–120
Ketamine/xylazine	200 mg kg^{-1} + 10 mg kg^{-1} i/p	Surgical anaesthesia	30–60	90–150
Pentobarbitone	50–90 mg kg^{-1} i/p	Immobilization/anaesthesia	30–60	120–180
Tiletamine/zolezepam	50–80 mg kg^{-1} i/p	Immobilization/anaesthesia	20–30	30–60
Tiletamine/zolezepam/ xylazine	30 mg kg^{-1} + 10 mg kg^{-1} i/p	Surgical anaesthesia	30	40–60
Urethane	1000–2000 mg kg^{-1}	Surgical anaesthesia	360–480	Non-recovery only

* Dose in ml kg^{-1} of a mixture of 1 part ‘Hypnorm’ plus 2 parts water for injection, and 1 part midazolam (5 mg ml^{-1} initial concentration).

Duration of anaesthesia and sleep time (loss of righting reflex) are provided only as a general guide, since considerable between animal variation occurs. For recommended techniques, see text.

D. Gerbils

1. Pre-anaesthetic medication

Initial restraint of gerbils for intraperitoneal administration of anaesthetics is reasonably simple, but young or particularly active individuals may be better anaesthetized using an anaesthetic chamber.

Information on the effects of sedative agents is limited in gerbils, but the following agents appear reasonably effective in this species (Table 7.8).

1. Fentanyl/fluanisone (Hypnorm, Janssen) (0.5–1.0 ml kg^{-1} i/p) provides sufficient analgesia for superficial procedures. Partial reversal is possible using nalbuphine (4 mg kg^{-1} i/p or s/c) or butorphanol (2 mg kg^{-1} i/p or s/c).
2. Diazepam (5 mg kg^{-1} i/p) or midazolam (5 mg kg^{-1} i/p) produces sedation but no analgesia.

Atropine (0.04 mg kg^{-1} s/c, i/m or i/p) can be administered to reduce salivary and bronchial secretions.

Dose rates for general anaesthesia given below should be reduced by 30–50% if one of the sedative drugs listed above has been administered.

TABLE 7.8
Sedatives, tranquillizers and other pre-anaesthetic medication for use in the gerbil

Drug	Dose rate	Comments
Acepromazine	3 mg kg^{-1} i/m	Light sedation
Atropine	0·04 mg kg^{-1} s/c	Anticholinergic
Diazepam	5 mg kg^{-1} i/m, i/p	Light sedation
Fentanyl/fluanisone (Hypnorm)	0·5–1.0 ml kg^{-1} i/m, i/p	Moderate sedation, moderate analgesia
Ketamine	100–200 mg kg^{-1} i/m	Heavy sedation, mild analgesia
Medetomidine	100–200 μg kg^{-1} i/p	Light to Heavy sedation, some analgesia
Midazolam	5 mg kg^{-1} i/m, i/p	Light/moderate sedation
Xylazine	2 mg kg^{-1} i/m	Light sedation, some analgesia

Considerable variation in effect occurs between different strains.

2. General anaesthesia

a. Injectable agents. Drugs are most conveniently administered by intraperitoneal injection in gerbils (Table 7.9).

1. The combination of fentanyl (0.05 mg kg^{-1} s/c) and metomidate (50 mg kg^{-1} s/c) appears most reliable in producing general anaesthesia in gerbils (Flecknell, 1983).

2. Fentanyl/fluanisone with midazolam or diazepam (0.3 ml kg^{-1} i/p Hypnorm, and diazepam 5 mg kg^{-1} i/p) is less satisfactory in gerbils than in other rodents, and only light anaesthesia may be produced. When using midazolam the components are mixed together with water for injection (see Appendix 3).

3. Ketamine (75 mg kg^{-1}) and medetomidine (0.5 mg kg^{-1}) mixed together and administered by intraperitoneal injection produce medium planes of anaesthesia in gerbils. The medetomidine may be reversed using atipamezole (1 mg kg^{-1} s/c or i/p).

4. As in hamsters, the use of pentobarbitone in gerbils is particularly hazardous and death often occurs if this drug is used to produce surgical anaesthesia (80 mg kg^{-1} i/p). Lower dose rates (60 mg kg^{-1} i/p) produce light anaesthesia which can be deepened using low concentrations of volatile anaesthetics (e.g. 0.5% halothane).

b. Inhalational agents. Halothane, isoflurane and methoxyflurane can all be used to provide effective and safe anaesthesia. Induction using an anaesthetic chamber is simple and convenient, followed by maintenance, if required, using a face-mask, as with other small rodents.

c. Endotracheal intubation. Endotracheal intubation requires the use of a purpose-built laryngoscope (Costa *et al.*, 1986).

d. Mechanical ventilation. Mechanical ventilation can be performed using a Harvard Rodent Ventilator (Appendix 7) — see Chapter 5.

3. Anaesthetic management

Like other small rodents, gerbils are especially susceptible to hypothermia and heating pads or lamps should be used to prevent this (see Chapter 4).

TABLE 7.9
Anaesthetic dose rates in the gerbil

Drug	Dose rate	Effect	Duration of anaesthesia (min)	Sleep time (min)
Alphaxalone/alphadolone	80–120 mg kg^{-1} i/p	Immobilization	–	60–90
Fentanyl/fluanisone (Hypnorm) + diazepam	0.3 ml kg^{-1} i/m or i/p + 5 mg kg^{-1} i/p	Surgical anaesthesia	20	60–90
Fentanyl/fluanisone (Hypnorm)/midazolam	8.0 ml kg^{-1} i/p*	Surgical anaesthesia	20	60–90
Ketamine/acepromazine	75 mg kg^{-1} + 3 mg kg^{-1} i/p	Immobilization	–	60–90
Ketamine/diazepam	50 mg kg^{-1} + 5 mg kg^{-1} i/p	Immobilization	–	30–60
Ketamine/medetomidine	75 mg kg^{-1} + 0.5 mg kg^{-1} i/p	Medium anaesthesia	20–30	90–120
Ketamine/xylazine	50 mg kg^{-1} + 2 mg kg^{-1} i/p	Immobilization	–	20–60
Metomidate/fentanyl	50 mg kg^{-1} + 0.05 mg kg^{-1} s/c	Surgical anaesthesia	45–90	180–240
Pentobarbitone	60–80 mg kg^{-1} i/p	Immobilization/anaesthesia	20	60–90
Tribromoethanol	250–300 mg kg^{-1} i/p	Surgical anaesthesia	15–30	30–90

* Dose in ml kg^{-1} of a mixture of 1 part 'Hypnorm' plus 2 parts water for injection, and 1 part midazolam (5 mg ml^{-1} initial concentration).

Duration of anaesthesia and sleep time (loss of righting reflex) are provided only as a general guide, since considerable between animal variation occurs. For recommended techniques, see text.

E. Guinea pigs

Guinea pigs are among the most difficult rodents in which to achieve safe and effective anaesthesia. Their response to many injectable anaesthetics is very variable and post-anaesthetic complications such as respiratory infections, digestive disturbances and generalized depression and inappetence are frequently seen. Many of these problems can be avoided by careful selection of anaesthetic agents and a high standard of intra-operative and post-operative nursing care.

1. Pre-anaesthetic medication

Guinea pigs are non-aggressive animals that are generally easy to handle and restrain. When frightened they run around their cage at high speed, making safe handling difficult. It is important to approach guinea pigs quietly and handle them gently but firmly. They should be picked up around the shoulders and thorax and the hindquarters supported as they are lifted clear of their cage. Intramuscular or intraperitoneal injection of anaesthetic agents can then be carried out. Pre-anaesthetic medication is therefore not usually required but, if an anaesthetic is to be administered by intravenous injection into an ear vein, initial sedation is advantageous. The following drugs can be used to produce sedation and restraint (Table 7.10).

1. Fentanyl/fluanisone (Hypnorm, Janssen) (1.0 ml kg^{-1} i/m or i/p) will produce restraint, sedation and sufficient analgesia for minor procedures such as skin biopsy.
2. Diazepam (5 mg kg^{-1} i/p) or midazolam (5 mg kg^{-1} i/p or i/m) produces heavy sedation and immobility, but no analgesia. The animal is easily roused by painful stimuli or other disturbances such as noise. These agents can be useful in providing sufficient sedation to allow local anaesthetic techniques to be used humanely.
3. Ketamine (100 mg kg^{-1} i/m) immobilizes guinea pigs but does not produce good analgesia.
4. Alphaxalone/alphadolone (40 mg kg^{-1} i/m) produces heavy sedation but requires a relatively large volume of drug to be injected (2–3 ml in an adult guinea pig).

Medetomidine and xylazine administered alone have very variable sedative effects in guinea pigs, but do potentiate the effects of other anaesthetic agents (see below).

Atropine (0.05 mg kg^{-1} s/c) should be administered to minimize the volume of bronchial and salivary secretions. It is particularly useful in

TABLE 7.10
Sedatives, tranquillizers and other pre-anaesthetic medication for use in the guinea pig.

Drug	Dose rate	Comments
Acepromazine	2·5–5 mg kg^{-1} i/m	Light to moderate sedation
Alphaxalone/alphadolone	40 mg kg^{-1} i/m, i/p	Heavy sedation, mild analgesia
Atropine	0·05 mg kg^{-1} s/c	Anticholinergic
Diazepam	5 mg kg^{-1} i/m, i/p	Heavy sedation
Fentanyl/droperidol (Innovar Vet)	0·44–0·8 ml kg^{-1} i/m	Sedation, analgesia
Fentanyl/fluanisone (Hypnorm)	1·0 ml kg^{-1} i/m, i/p	Moderate sedation, moderate analgesia
Ketamine	100 mg kg^{-1} i/m	Heavy sedation, light analgesia
Midazolam	5 mg kg^{-1} i/m, i/p	Heavy sedation

Considerable variation in effect occurs between different strains.

guinea pigs because of their relatively narrow airways, which are prone to obstruction.

The dose rates of drugs listed below apply to guinea pigs that have received no pre-anaesthetic medication unless otherwise stated. The dosages of general anaesthetic agents should be reduced by 30–50% if one of the drugs listed above has been administered.

2. *General anaesthesia*

a. Injectable agents. Intravenous administration of anaesthetic drugs is difficult in guinea pigs and drugs are usually administered by the intraperitoneal, subcutaneous or intramuscular routes. The animals should be carefully weighed and dose rates calculated accurately. Anaesthetic dose rates are summarized in Table 7.11.

1. The anaesthetic combination of choice is fentanyl/fluanisone (Hypnorm, Janssen) together with midazolam or diazepam (1.0 ml kg^{-1} i/p Hypnorm, and diazepam 2.5 mg kg^{-1} i/p). When using midazolam the components are mixed together with water for injection (see Appendix 3). These combinations provide good surgical anaesthesia lasting about 45 minutes. (Green, 1975; Flecknell and Mitchell, 1984). If a longer period of anaesthesia is required, further doses of Hypnorm can be given

TABLE 7.11
Injectable anaesthetic dose rates in the guinea pig

Drug	Dose rate	Effect	Duration of anaesthesia (min)	Sleep time (min)
Alphaxalone/alphadolone	40 mg kg^{-1} i/p	Immobilization	–	90–120
Alpha-chloralose	70 mg kg^{-1} i/p	Light to medium anaesthesia	180–600	Non-recovery only
Fentanyl/fluanisone (Hypnorm) + diazepam	1 ml kg^{-1} i/m or i/p + 2·5 mg kg^{-1} i/p	Surgical anaesthesia	45–60	120–180
Fentanyl/fluanisone (Hypnorm)/midazolam	8·0 ml kg^{-1} i/p*	Surgical anaesthesia	45–60	120–180
Ketamine/acepromazine	125 mg kg^{-1} + 5 mg kg^{-1} i/m	Immobilization/anaesthesia	45–120	90–180
Ketamine/diazepam	100 mg kg^{-1} + 5 mg kg^{-1} i/m	Immobilization/anaesthesia	30–45	90–120
Ketamine/medetomidine	40 mg kg^{-1} + 0·5 mg kg^{-1} i/p	Moderate anaesthesia	30–40	90–120
Ketamine/xylazine	40 mg kg^{-1} + 5 mg kg^{-1} i/p	Surgical anaesthesia	30	90–120
Methohexitone	31 mg kg^{-1} i/p	Immobilization	–	20
Pentobarbitone	37 mg kg^{-1} i/p	Surgical anaesthesia	60–90	240–300
Tiletamine/zolezepam	40–60 mg kg^{-1} i/m	Immobilization	–	70–160
Urethane	1500 mg kg^{-1} i/v, i/p	Surgical anaesthesia	300–480	Non-recovery only

* Dose in ml kg^{-1} of a mixture of 1 part 'Hypnorm' plus 2 parts water for injection, and 1 part midazolam (5 mg ml^{-1} initial concentration).

Duration of anaesthesia and sleep time (loss of righting reflex) are provided only as a general guide, since considerable between animal variation occurs. For recommended techniques, see text.

(approximately 0.5 ml kg^{-1} i/m every 20–30 minutes). Following the completion of surgery the anaesthesia can be partially reversed using nalbuphine (1 mg kg^{-1} i/p or s/c), butorphanol (1 mg kg^{-1} i/p or s/c) or buprenorphine (0.01 mg kg^{-1} i/v or 0.05 mg kg^{-1} i/p).

2. An effective alternative is to administer ketamine (40 mg kg^{-1} i/m) and xylazine (5 mg kg^{-1} s/c). This combination provides about 30 minutes of surgical anaesthesia, although the degree of analgesia may be insufficient for major surgery in some animals (D'Alleine and Mann, 1982; Hart *et al.*, 1984; Barzago *et al.*, 1994). Ketamine (40 mg kg^{-1} i/p) and medetomidine (0.5 mg kg^{-1} i/p) produce moderate anaesthesia (Nevalainen *et al.*, 1989). Anaesthesia can be partially reversed using atipamezole (1 mg kg^{-1} s/c). Either of these ketamine combinations can be administered as a single injection by the intraperitoneal route.

3. Alphaxalone/alphadolone produces only light surgical anaesthesia even when administered by the intravenous route. If additional anaesthetic is administered, severe respiratory depression frequently ensues.

4. If pentobarbitone is to be used, it is best administered at a dose of 25 mg kg^{-1} i/p to sedate and immobilize the animal, and anaesthesia should then be deepened using a volatile agent such as methoxyflurane. Use of the higher dose rates of pentobarbitone (37 mg kg^{-1} i/p), which are needed to produce surgical anaesthesia, are frequently associated with an unacceptably high mortality rate.

b. Inhalational agents. Induction of anaesthesia can be either by use of an anaesthetic chamber or by administration via a small face-mask. Following induction, it is usually most convenient to maintain anaesthesia using a face-mask, since endotracheal intubation is an extremely difficult technique to carry out in guinea pigs. If it is necessary to intubate the animal, use can be made of a purpose-made laryngoscope (Costa *et al.*, 1986; Blouin and Cormier, 1987). Alternatively, the larynx can be visualized by transilluminating the neck and placing an otoscope speculum in the oral cavity (see Chapter 3). A soft-tipped wire introducer is then used to insert a catheter into the larynx.

Methoxyflurane is an effective volatile anaesthetic in guinea pigs, is non-irritant, and has a wide margin of safety, although as with other species, induction is slow (4–5 minutes). Halothane can be used successfully, but can produce profound hypotension even at normal maintenance concentrations. Isoflurane can also be used successfully, but hypotension may also be produced.

Ether is unsuitable for use in guinea pigs because it is highly irritant to the respiratory tract, producing increased bronchial secretions that tend to occlude the narrow airways. In addition, bronchospasm may be produced during induction of anaesthesia with ether.

3. Local anaesthetic techniques

Intrathecal (spinal) anaesthesia using local anaesthetic agents has been described in the guinea pig (Thomasson *et al.*, 1974), and this technique coupled with the use of sedatives or low doses of other anaesthetic agents for restraint may be useful in some circumstances.

4. Anaesthetic management

Care must be taken to prevent the development of hypothermia, using the methods described in Chapter 4. Although high standards of post-operative care are required for all species, they are particularly important in guinea pigs. Post-operative recovery is aided by administering 10–15 ml of warmed dextrose-saline (0.18% saline, 4% dextrose) s/c to correct any fluid deficit. A warm (25–30°C) recovery area should be provided, and the animal given additional subcutaneous fluid for the next few days if its appetite is depressed.

II. RABBITS

Rabbits are easily stressed by inexpert pre-operative handling and during induction with volatile anaesthetic agents. The combined effects of stress and anaesthesia can result in cardiac and respiratory arrest. In addition, the frequent presence of pre-existing lung damage caused by infection with *Pasteurella multocida* may result in respiratory failure during the period of anaesthesia. Recovery from anaesthesia is often slow, particularly following the use of barbiturates, and the prolonged inappetence that is a frequent post-operative complication can result in gastrointestinal disturbances.

The incidence of these potentially serious problems can be minimized by careful selection of the anaesthetic regimen, avoidance of stress both pre-operatively and post-operatively, and by maintaining high standards of intra-operative and post-operative care. When examining rabbits before anaesthesia, particular attention should be given to the eyes and nose — the presence of any abnormal discharges is strongly suggestive of respiratory tract infection. The forelegs should also be inspected, as nasal discharge may become encrusted on the fur during grooming. It is unnecessary to withhold food and water prior to induction of anaesthesia since vomiting during induction or recovery does not occur in rabbits. The large caecum in this species and the occurrence of coprophagy also makes pre-operative fasting relatively ineffective in reducing the mass of abdominal viscera prior to abdominal surgery.

A. Pre-anaesthetic medication

Because rabbits are so easily stressed, whenever possible a tranquillizer or sedative should be administered while the animal is still in its familiar surroundings. The rabbit can then be transported to the operating theatre when sedated, so minimizing the stress caused by such manipulations. A wide variety of tranquillizers can be used successfully in rabbits, and these are listed in order of preference below (Table 7.12).

1. Fentanyl/fluanisone (Hypnorm, Janssen) (0.2–0.5 mg kg^{-1} i/m). At the lower dose rate sedation and some analgesia is produced (Green, 1975). The higher dose rate produces sufficient analgesia to enable procedures such as draining and cleaning of subcutaneous abscesses to be carried out. Occasionally, marked respiratory depression is seen when the drug is administered at the higher dose rate. If this produces marked cyanosis, oxygen should be administered and the fentanyl component of the mixture reversed by the administration of nalbuphine (1 mg kg^{-1} s/c or i/p), butorphanol (1 mg kg s/c or i/p) or buprenorphine (0.05 mg kg^{-1} s/c or i/p) (Flecknell *et al.*, 1989a).
2. Medetomidine (0.25 mg kg^{-1} i/m) also produces safe and effective sedation. At higher doses (0.5 mg kg^{-1} i/m) animals lose their righting reflex. Only minimal analgesia is produced when medetomidine is administered alone. Sedation is completely reversed by administration of atipamezole (0.2 mg kg^{-1} i/v, 1.0 mg kg^{-1} s/c).
3. Xylazine (2–5 mg kg^{-1} i/m) produces light to heavy sedation but appears to have little analgesic action when used alone in rabbits. Sedation can be reversed using atipamezole (1.0 mg kg^{-1} s/c) or yohimbine (0.2 mg kg^{-1} i/v) (Lipman *et al.*, 1987).
4. Acepromazine (1 mg kg^{-1} i/m) produces moderate sedation but has no analgesic action. Combining acepromazine (0.5 mg kg^{-1}) and butorphanol (0.5 mg kg^{-1}) produces good sedation combined with moderate analgesia (COMPMED, personal communication 1995).
5. Diazepam or midazolam (0.5–2 mg kg^{-1} i/v, i/m or i/p) produces good sedation, but neither drug has any analgesic action. Sedation can be

TABLE 7·12
Sedatives, tranquillizers and other pre-anaesthetic medication for use in the rabbit

Drug	Dose rate	Comments
Acepromazine	1 mg kg^{-1} i/m	Moderate sedation
Alphaxalone/alphadolone	9–12 mg kg^{-1} i/m	Moderate to heavy sedation, little analgesia
Atropine	0·05 mg kg^{-1} i/m	Very short-acting in some rabbits
Acepromazine + butorphanol	1 mg kg^{-1} + 1 mg kg^{-1} i/m	Moderate to heavy sedation, moderate analgesia
Diazepam	0·5–2 mg kg^{-1} i/v, i/m, i/p	Light to moderate sedation
Glycopyrrolate	0.01 mg kg^{-1} i/v, 0·1 mg kg^{-1} s/c, i/m	Anticholinergic
Fentanyl/droperidol (Innovar Vet)	0·22 ml kg^{-1} i/m	Immobilization, analgesia
Fentanyl/fluanisone (Hypnorm)	0·2–0·5 ml kg^{-1} i/m	Light to heavy sedation, light to deep analgesia
Ketamine	25–50 mg kg^{-1} i/m	Moderate to heavy sedation, mild analgesia
Medetomidine	0·1–0·5 mg kg^{-1} i/m, s/c	Light to heavy sedation, some analgesia
Midazolam	0·5–2 mg kg^{-1} i/v, i/m, i/p	Light to moderate sedation
Xylazine	2–5 mg kg^{-1} i/m	Light to moderate sedation, some analgesia

Considerable variation in effect occurs between different strains.

reversed using flumazenil (0.01–0.1 mg kg^{-1} i/v), but the animal may become sedated again a few hours later.

6. Ketamine (25–50 mg kg^{-1} i/m) produces heavy sedation. As with other species, the degree of muscle relaxation is poor and the level of analgesia is insufficient for even superficial surgery.

7. Alphaxalone/alphadolone (9–12 mg kg^{-1} i/m) produces moderate to heavy sedation, the effects varying considerably between individuals. The degree of analgesia is generally low.

8. Glycopyrrolate (0.01 mg kg^{-1} i/v, 0.1 mg kg^{-1} s/c or i/m) (Olson *et al.*, 1993) may be administered to reduce salivary and bronchial secretions and protect the heart from vagal inhibition. Atropine is relatively ineffective in many strains of rabbits because of high levels of atropinase in the liver, and repeated doses are usually required if this drug is used.

In addition to reducing the stress caused by induction of anaesthesia and easing handling and restraint, the use of appropriate pre-anaesthetic medication reduces the amount of other anaesthetic agents required to produce general anaesthesia. The dose rates quoted in Table 7.13 apply to rabbits that have received no pre-anaesthetic medication unless otherwise stated. The dose rates for intravenous induction agents can be reduced by 30–50% if one of the drugs listed above has been administered.

B. General anaesthesia

1. Injectable agents

Intravenous injection can be carried out relatively easily using the marginal ear vein. Injection is made even easier if the ear is treated with local anaesthetic cream (Emla, Astra) 45 minues prior to injection (see Chapter 2). Anaesthetic dose rates are summarized in Table 7.13.

1. The drug combination of choice for routine anaesthesia of rabbits is fentanyl/fluanisone (Hypnorm) (0.3 ml kg^{-1} i/m) and midazolam or diazepam (2 mg kg^{-1} i/m, i/v or i/p) (Flecknell *et al.*, 1983; Flecknell and Mitchell, 1984). This combination provides good surgical anaesthesia with excellent muscle relaxation for about 20–40 minutes. It is recommended that fentanyl/fluanisone be administered first, and 10–15 minutes allowed to elapse so that the animal becomes sedated. The rabbit can then be transferred to the operating theatre or procedure room without distress. Because marked analgesia is produced, the animal will be unresponsive to the pain of intravenous injection, and the vasodilation caused by Hypnorm

TABLE 7·13
Anaesthetic dose rates in the rabbit

Drug	Dose rate	Effect	Duration of anaesthesia (min)	Sleep time (min)
Alphaxalone/alphadolone	6–9 mg kg^{-1} i/v	Light anaesthesia	5–10	10–20
Alpha-chloralose	80–100 mg kg^{-1} i/v	Light to surgical anaesthesia	360–600	Non-recovery only
Etorphine/methotrimeprazine (Immobilon SA)	0·025–0·05 ml kg^{-1} i/m	Immobilization, analgesia	60 (analgesia)	120–240
Etorphine/methotrimeprazine (Immobilon SA) + midazolam	0·05 ml kg^{-1} i/m + 1 mg kg^{-1} i/v	Surgical anaesthesia (severe respiratory depression, see text)	50–100	180–240
Fentanyl/fluanisone (Hypnorm) + diazepam	0·3 ml kg^{-1} i/m + 1–2 mg kg^{-1} i/v, i/m or i/p	Surgical anaesthesia	20–40	60–120
Fentanyl/fluanisone (Hypnorm) + midazolam	0·3 ml kg^{-1} i/m + 1–2 mg kg^{-1} i/v or i/p	Surgical anaesthesia	20–40	60–120
Fentanyl + medetomidine	8 μg kg^{-1} i/v + 330 μg kg^{-1} i/v		30–40	90–180
Ketamine/acepromazine	50 mg kg^{-1} i/m + 1 mg kg^{-1} i/m	Surgical anaesthesia	20–30	60–90
Ketamine/diazepam	25 mg kg^{-1} i/m + 5 mg kg^{-1} i/m	Surgical anaesthesia	20–30	60–90
Ketamine/medetomidine	25 mg kg^{-1} i/m + 0·5 mg kg^{-1} i/m	Surgical anaesthesia	30–40	120–240

Drug	Dose rate	Effect	Duration of anaesthesia (min)	Sleep time (min)
Ketamine/xylazine	35 mg kg^{-1} i/m + 5 mg kg^{-1} i/m	Surgical anaesthesia	25–40	60–120
	10 mg kg^{-1} i/v + 3 mg kg^{-1} i/v	Surgical anaesthesia	20–30	60–90
Ketamine/xylazine/acepromazine	35 mg kg^{-1} i/m + 5 mg kg^{-1} i/m + 1·0 mg kg^{-1} i/m, s/c	Surgical anaesthesia	45–75	100–150
Ketamine/xylazine/butorphanol	35 mg kg^{-1} i/m + 5 mg kg^{-1} i/m + 0·1 mg kg^{-1} i/m	Surgical anaesthesia	60–90	120–180
Methohexitone	10–15 mg kg^{-1} i/v	Surgical anaesthesia	4–5	5–10
Pentobarbitone	30–45 mg kg^{-1} i/v	Light to medium anaesthesia	20–30	60–120
Propofol	10 mg kg^{-1} i/v	Light anaesthesia	5–10	10–15
Thiopentone	30 mg kg^{-1} i/v	Surgical anaesthesia	5–10	10–15
Urethane	1000–2000 mg kg^{-1} i/v	Surgical anaesthesia	360–480	Non-recovery only

Duration of anaesthesia and sleep time (loss of righting reflex) are provided only as a general guide, since considerable between animal variation occurs. For recommended techniques, see text.

also aids placement of an intravenous needle or over-the-needle catheter. Midazolam or diazepam can then be administered to effect to produce loss of consciousness and relaxation.

Longer periods of anaesthesia can be achieved by the administration of additional doses of Hypnorm (approximately 0.1 ml kg^{-1} i/v every 30–40 minutes). This is best achieved by diluting the commercial preparation 1:10 with water for injection. Use of saline results in precipitation of one of the components of Hypnorm. If anaesthesia of several hours' duration is required, it is preferable to administer fentanyl (30–100 μg kg^{-1} h^{-1}) alone, to avoid undue accumulation of the fluanisone component of Hypnorm.

Following the completion of surgery the anaesthesia can be reversed using nalbuphine (0.1 mg kg^{-1} i/v) or buprenorphine (0.01 mg kg^{-1} i/v).

2. The most useful alternative is to administer ketamine (25 mg kg^{-1} i/m) and medetomidine (0.5 mg kg^{-1} i/m) (Nevalainen *et al.*, 1989), or ketamine (35 mg kg^{-1} i/m) and xylazine (5 mg kg^{-1} i/m), which both provide about 30 minutes of surgical anaesthesia (White and Holmes, 1976; Lipman *et al.*, 1990). The duration of anaesthesia can be prolonged by the addition of butorphanol (0.1 mg/kg) to the ketamine/xylazine regimen, resulting in approximately 80 minutes of surgical anaesthesia (Marini *et al.*, 1992). The degree of cardiovascular and respiratory depression is generally greater than that seen with the Hypnorm/diazepam or Hypnorm/midazolam mixture described above. Anaesthesia can be partially reversed using atipamezole (1 mg kg^{-1} s/c or i/v) or yohimbine (Lipman *et al.*, 1987).

3. Fentanyl (8 μg kg^{-1}) and medetomidine (330 μg kg^{-1}) administered in combination by intravenous injection produce good surgical anaesthesia in rabbits, but some animals may make spontaneous movements in response to non-painful stimuli, and in general this combination is less satisfactory than it is in the rat. An advantage of the combination is that it can be completely reversed using atipamezole (1 mg kg^{-1} i/v) and nalbuphine (1 mg kg^{-1} i/v). It is important that the rabbit is placed in a suitable recovery cage immediately anaesthesia is reversed, as full recovery may occur in less than 1 minute.

4. Thiopentone (30 mg kg^{-1} i/v, using a 1.25% solution) or methohexitone (10–15 mg kg^{-1} i/v, using a 1% solution) can be administered to produce 5–10 minutes of anaesthesia.

5. Propofol (10 mg kg^{-1} i/v) is less effective in the rabbit than other species, and only light anaesthesia is produced at this dose rate. Higher doses (15–20 mg kg^{-1}) cause respiratory arrest. Attempts to produce prolonged anaesthesia in rabbits with propofol have been less successful

than in other species (Glen, 1980; Blake *et al.*, 1988; Ko *et al.*, 1992; Aeschbacher and Webb, 1993a, b).

6. Similar effects to propofol are seen with alphaxalone/alphadolone (6–9 mg kg^{-1} i/v) which produces light general anaesthesia but, at the higher dose rates necessary to produce medium or deep surgical anaesthesia, causes sudden apnoea (Green *et al.*, 1978). This may rapidly be followed by cardiac arrest, so this agent is not recommended in rabbits. This drug is perhaps best used at the lower dose rate (6 mg kg^{-1}) prior to increasing the depth of anaesthesia with a volatile anaesthetic.

7. Pentobarbitone (30–45 mg kg^{-1} i/v), if it must be used, should be diluted to provide a 30 mg ml^{-1} solution and administered slowly to effect. Considerable skill and extensive practical experience is required to use this drug effectively in the rabbit. Respiratory arrest frequently occurs before the onset of surgical anaesthesia and because of the consequent high mortality the drug is best avoided in this species (Flecknell *et al.*, 1983; Peeters *et al.*, 1988).

2. *Inhalational agents*

It is possible to induce anaesthesia in rabbits using only inhalational agents, but such inductions are usually stressful both for the patient and the anaesthetist. Induction is made more hazardous by the occurrence of a breath-holding response in the rabbit (Flecknell *et al.*, 1996): exposure even to low concentrations of halothane or isoflurane, delivered either by face-mask or via an anaesthetic chamber, caused apnoea for periods of up to 2 minutes. Administration of medetomidine, midazolam or acepromazine did not block this response. It is therefore generally preferable to induce anaesthesia with an injectable agent and maintain the rabbit on an inhalational anaesthetic. If induction with an inhalational agent is required, the animal should be observed closely for episodes of breath-holding. Face-mask induction is preferable, since it allows better control of administration, and enables emergency resuscitation to be undertaken if required. Anaesthetic chambers have the added disadvantage that involuntary excitement during induction and consequent injury may occur.

Provided that the problems of induction mentioned above are noted, halothane and isoflurane can both be used to produce either short-term or long-term anaesthesia in rabbits. Both agents should always be administered using a calibrated vaporizer.

Methoxyflurane is also a safe and effective anaesthetic in the rabbit. Although this agent produces some cardiovascular and respiratory depression, this is less than that seen at comparable depths of anaesthesia with isoflurane or halothane.

Ether is an unsuitable agent for use in rabbits — its irritant nature can result in laryngospasm if used for induction and it frequently exacerbates pre-existing respiratory disease. Its irritant properties also result in profuse bronchial and salivary secretions. It is also explosive when mixed with oxygen or air and this can make it a serious safety hazard.

3. *Endotracheal intubation*

Endotracheal intubation can be carried out by a number of different techniques, using a 3–4 mm endotracheal tube (see Chapter 3).

The most useful breathing circuits for rabbits are the Ayre's T-piece and the Bain coaxial circuit (Chapter 3).

III. CATS

A. Pre-anaesthetic medication

The majority of cats respond well to firm but gentle physical restraint, enabling intravenous administration of anaesthetic drugs into the cephalic vein on the forelimb. Prior application of local anaesthetic cream (EMLA) can prevent any struggling in response to intravenous injection (Flecknell *et al.*, 1990a), see Chapter 1. If an experienced assistant is not available, it may be more convenient to administer drugs by the subcutaneous or intramuscular routes. Some pre-anaesthetic agents such as medetomidine can produce complete relaxation and loss of consciousness at higher doses (Table 7.14).

Cats should be starved for 12 hours prior to induction of anaesthesia to minimize the risk of vomiting during induction or during the recovery period. In an emergency, if fasting has not been possible, then medetomidine can be administered since this drug causes vomiting prior to producing sedation. The drugs listed below can all be used to produce sedation and will ease handling and restraint. Some of the agents can produce a sufficient depth of anaesthesia to enable minor procedures to be carried out. An extensive review of feline anaesthesia is given by Hall and Taylor (1994).

1. Ketamine (5–20 mg kg^{-1} i/m) produces moderate analgesia and sedation and the higher dose rate will immobilize a cat for about 30–45 minutes. Skeletal muscle tone is increased, making minor manipulations difficult but, since pharyngeal and laryngeal reflexes are maintained, the drug is particularly useful if the animal has not fasted. The palpebral

TABLE 7.14
Sedatives, tranquillizers and other pre-anaesthetic medication for use in the cat

Drug	Dose rate	Comments
Acepromazine	0·05–0·2 mg kg^{-1} i/m	Light to moderate sedation
Acepromazine + buprenorphine	0·05 mg kg^{-1} i/m 0·01 mg kg^{-1} i/m	Heavy sedation, immobilization
Acepromazine + morphine	0·05 mg kg^{-1} i/m 0·1 mg kg^{-1} i/m	Heavy sedation, immobilization
Alphaxalone/alphadolone	9 mg kg^{-1} i/m	Moderate to heavy sedation
Atropine	0·05 mg kg^{-1} i/m, s/c	Anticholinergic
Glycopyrrolate	0·01 mg kg^{-1} i/v 0·05 mg kg^{-1} i/m	Anticholinergic
Ketamine	5–20 mg kg^{-1} i/m, 10–20 mg kg^{-1} per os	Moderate to heavy sedation, moderate analgesia
Medetomidine	50–150 μg kg^{-1} i/m or s/c	Light to heavy sedation, moderate analgesia
Pethidine	3–5 mg kg^{-1} i/m or s/c	Light sedation, mild analgesia
Xylazine	1–2 mg kg^{-1} i/m or s/c	Light to moderate sedation, moderate analgesia

and corneal blink reflexes are lost and if prolonged anaesthesia is anticipated the eyes should be filled with a bland ophthalmic ointment to prevent damage to the cornea through desiccation. Although the volume of injection is small, the low pH of the solution makes intramuscular injection painful.

2. Medetomidine (50–150 μg kg^{-1} i/m or s/c) produces light to heavy sedation, lasting 60–90 minutes. Sedation can be completely and rapidly reversed with atipamezole (100–600 μg kg^{-1} s/c, i/m or i/v). Cats commonly vomit during the onset of sedation. The degree of analgesia is sufficient for minor procedures, such as percutaneous passage of a large-gauge over-the-needle catheter, but insufficient for surgical procedures.

3. Xylazine (1–2 mg kg^{-1} i/m or s/c) produces good sedation lasting 30–40 minutes. Vomiting commonly occurs during the onset of sedation. The degree of sedation is sufficient to enable minor manipulation to be carried

out, but the analgesic effects of the drug are variable. Reversal of sedation can be achieved by administration of atipamezole (100–600 μg kg^{-1} s/c, i/m or i/v).

4. Acepromazine (0.05–0.2 mg kg^{-1} i/m) tranquillizes cats prior to induction of anaesthesia and eliminates the excitement associated with recovery from barbiturate anaesthesia.

5. Alphaxalone/alphadolone (9 mg kg^{-1} i/m) produces heavy sedation sufficient to carry out minor, non-painful, manipulative procedures. To be effective, this agent must be administered into a muscle mass, rather than into intramuscular fascia. Full surgical anaesthesia can be produced by administering additional drug by the intravenous route.

6. Pethidine (3–5 mg kg^{-1} i/m or s/c) will make some cats more tractable and provides some analgesia, although the sedative effects in certain individuals appears minimal.

Atropine (0.05 mg kg^{-1} i/m) or glycopyrrolate (0.01 mg kg^{-1} i/v, 0.05 mg kg^{-1} i/m) can be administered prior to induction of anaesthesia to reduce salivation and protect the heart from vagally mediated bradycardia.

The dose rates of anaesthetic drugs listed below and in Table 7.15 can be reduced by 30–50% if one of the drugs (1–6) listed above has been administered.

B. General anaesthesia

1. Injectable agents

1. Propofol (5–8 mg kg^{-1} i/v) will provide about 10 minutes of surgical anaesthesia. Induction and recovery are smooth and rapid, and incremental doses can be used to prolong anaesthesia without unduly prolonging recovery times (Glen, 1980; Brearley *et al.*, 1988).

2. Alphaxalone/alphadolone (9 mg kg^{-1} i/v) provides about 10 minutes of surgical anaesthesia. Incremental injections of approximately 3 mg kg^{-1} can be given to prolong anaesthesia. Alternatively, the drug can be given by continuous infusion at a rate of about 0.2 mg kg^{-1} min^{-1}.

3. Ketamine (20–30 mg kg^{-1} i/m) can be used to provide sufficient analgesia and restraint for minor surgical procedures. Improved muscle relaxation can be achieved by the concurrent administration of medetomidine or xylazine. The dosage of ketamine required can be reduced to

7 mg kg^{-1} i/m ketamine and medetomidine (80 μg kg^{-1} i/m or s/c) or 22 mg kg^{-1} i/m ketamine and xylazine (1.1 mg kg^{-1} i/m or s/c). Partial reversal of anaesthesia can be achieved by administration of atipamezole (0.3–0.5 mg/kg i/v or s/c) (Verstegen *et al.*, 1989, 1990, 1991a, b).

Ketamine (20 mg kg^{-1}) combined with acepromazine (0.11 mg kg^{-1}) provides light to moderate surgical anaesthesia (Colby and Sanford, 1981; Ingwersen *et al.*, 1988). The degree of analgesia may be improved by pre-anaesthetic administration of an opioid such as butorphanol (0.4 mg kg^{-1} s/c) to increase the degree of intra-operative analgesia (Tranquilli *et al.*, 1988). As with other combinations, addition of a tranquillizer reduces the muscle rigidity associated with ketamine alone and appears to produce unconsciousness and a state more resembling conventional general anaesthesia, although the eyes remain open with a dilated pupil.

4. Thiopentone (10–15 mg kg^{-1} i/v) (1.25% solution) (Middleton *et al.*, 1982) and methohexitone (4–8 mg kg^{-1}) (1% solution) will each produce about 5 minutes of light anaesthesia.

5. Pentobarbitone (20–30 mg kg^{-1} i/v) produces 30–90 minutes light to moderate surgical anaesthesia. In order to avoid involuntary excitement during induction, half of the calculated dose should be administered rapidly, followed by the remainder more slowly, to effect. Pentobarbitone has a relatively slow onset of action, so that administration of the remaining dose should generally take around 5–6 minutes. Too-rapid injection is frequently associated with apnoea and severe cardiovascular depression. Alternative routes of administration (intraperitoneal, oral, intrathoracic and subcutaneous) have been described but these are generally unsatisfactory and not recommended (Clifford and Soma, 1969). Pentobarbitone has a narrow safety margin in the cat, as in other species, and dose rates of 72 mg kg^{-1} have been reported to be lethal (Clifford and Soma, 1969). Recovery can be prolonged, especially if the animal is allowed to become hypothermic, and may be associated with excitement. Cats may remain ataxic and sedated for 8–24 hours.

6. Alpha-chloralose (70 mg kg^{-1} i/p or 60 mg kg^{-1} i/v) is suitable for prolonged, non-recovery procedures (see Chapter 5). In the cat, medium planes of surgical anaesthesia are produced in most animals. As in other species, induction is slow and may be associated with excitement, so it is best to administer a short-acting anaesthetic such as propofol before administration of chloralose.

TABLE 7.15
Anaesthetic dose rates in the cat

Drug	Dose rate	Effect	Duration of anaesthesia (min)	Sleep time (min)
Alphaxalone/alphadolone	9–12 mg kg^{-1} i/v, 18 mg kg^{-1} i/m	Surgical anaesthesia	10–15	45–120
Alpha-chloralose	70 mg kg^{-1} i/p, 60 mg kg^{-1} i/v	Light to medium anaesthesia	180–720	Non-recovery only
Fentanyl/metomidate	0·02 mg kg^{-1} i/m + 20 mg kg^{-1} i/m	Surgical anaesthesia	–	300
Ketamine/acepromazine	20 mg kg^{-1} i/m + 0·11 mg kg^{-1} i/m	Surgical anaesthesia	20–30	180–240
Ketamine/medetomidine	7 mg kg^{-1} i/m + 80 μg kg^{-1} i/m	Surgical anaesthesia	30–40	180–240
Ketamine/midazolam	10 mg kg^{-1} i/m + 0·2 mg kg^{-1} i/m	Surgical anaesthesia	20–30	180–240
Ketamine/promazine	15 mg kg^{-1} i/m + 1·12 mg kg^{-1} i/m	Surgical anaesthesia	20–30	180–240
Ketamine/xylazine	22 mg kg^{-1} i/m + 1·1 mg kg^{-1} i/m	Surgical anaesthesia	20–30	180–240
Methohexitone	4–8 mg kg^{-1} i/v	Surgical anaesthesia	5–6	60–90
Pentobarbitone	20–30 mg kg^{-1} i/v	Surgical anaesthesia	60–90	240–480
Propanidid	8–16 mg kg^{-1} i/v	Surgical anaesthesia	4–6	20–30
Propofol	5–8 mg kg^{-1} i/v	Surgical anaesthesia	5–10	20
Thiamylal	12–18 mg kg^{-1} i/v	Surgical anaesthesia	10–15	60–120

Drug	Dose rate	Effect	Duration of anaesthesia (min)	Sleep time (min)
Thiopentone	10–15 mg kg^{-1} i/v	Surgical anaesthesia	5–10	60–120
Tiletamine/zolezepam	7·5 mg kg^{-1} i/m + 7·5 mg kg^{-1} i/m	Surgical anaesthesia	20–40	
Urethane	750 mg kg^{-1} i/v, 1500 mg kg^{-1} i/p	Surgical anaesthesia	360–480	Non-recovery only

Duration of anaesthesia and sleep time (loss of righting reflex) are provided only as a general guide, since considerable between-animal variation occurs. For recommended techniques, see text.

2. *Inhalational agents*

It is preferable to induce anaesthesia using an injectable agent and then maintain anaesthesia using an inhalational agent, since cats often resent the process of face-mask induction. Prior administration of a sedative/tranquillizer coupled with expert handling may, however, enable smooth induction of anaesthesia by this method. An alternative, which avoids the need for firm physical restraint, is to use an anaesthetic chamber.

Laryngospasm during induction with volatile anaesthetics may occur and the incidence of this problem may be reduced by increasing anaesthetic concentrations gradually.

Following induction of anaesthesia, either with an injectable or volatile anaesthetic, the cat should be intubated using an uncuffed, 3–4 mm endotracheal tube. Endotracheal intubation in the cat can be carried out under direct vision using a paediatric laryngoscope blade. Care must be taken to spray the larynx with 2% lignocaine before attempting intubation, to help prevent the development of laryngospasm. In the UK, the most recent formulation of lignocaine (Astra) has been associated with the occurrence of laryngeal oedema and it is advisable to check the suitability of any locally available product before use (see Chapter 3). In most instances, some spasm occurs immediately following spraying of the larynx, but this passes rapidly and intubation can then be successfully achieved. Alternatively suxamethonium (1.0 mg kg^{-1} i/v) can be administered following induction of anaesthesia and the animal ventilated for a short period following intubation. Under these circumstances ventilation can be carried out using a face-mask if difficulties are experienced during intubation.

Following intubation, it is preferable to attach the animal to an Ayre's T-piece or Bain circuit, since expiratory resistance and equipment dead space are minimal when using these circuits.

Methoxyflurane, halothane, isoflurane and enflurane can all be used for maintenance of anaesthesia, but ether is best avoided because of its irritant nature.

IV. DOGS

A. Pre-anaesthetic medication

Dogs respond positively to human contact, and if the animal's regular handler is present restraint will rarely be a problem. It is often preferable, however, to administer pre-anaesthetic medication to dogs, to ease handling, to ensure a smooth and stress-free induction of anaesthesia and also to provide a quiet and gradual recovery. Dogs should be fasted for 12 hours prior to induction of anaesthesia. Intravenous injection for induction of anaesthesia is easy to carry out, particularly if the skin has been anaesthetized by prior application of local anaesthetic cream (EMLA) (see Chapter 2), and a sedative or tranquillizer has been administered.

The following drugs can be used for pre-anaesthetic medication and are listed in order of preference (Table 7.16).

1. Medetomidine (0.1–0.8 mg kg^{-1} i/m, s/c or i/v) produces light to heavy sedation and at higher dose rates the dog is completely immobilized, enabling minor procedures to be carried out. The degree of analgesia is insufficient for anything other than superficial surgical procedures. Sedation can be reversed completely and rapidly by administration of atipamezole (50–400 μg kg^{-1} i/m or i/v).
2. Buprenorphine (0.009 mg kg^{-1} i/m) and acepromazine (0.07 mg kg^{-1} i/m) produce moderate or heavy sedation, enabling minor procedures such as radiography to be undertaken easily (Taylor and Herrtage 1986).

TABLE 7.16
Sedatives, tranquillizers and other pre-anaesthetic medication for use in the dog

Drug	Dose rate	Comments
Acepromazine	0·1–0·25 mg kg^{-1} i/m	Light to moderate sedation
Acepromazine/ buprenorphine	0·07 mg kg^{-1} i/m + 0·009 mg kg^{-1} i/m	Heavy sedation, immobilization, some analgesia
Atropine	0·05 mg kg^{-1} s/c or i/m	Anticholinergic
Etorphine methotrimeprazine 'Immobilon SA'	0·5 ml/4 kg i/m	Immobilization/analgesia
Glycopyrrolate	0·01 mg kg^{-1} i/v	Anticholinergic
Fentanyl/droperidol (Innovar Vet)	0·1–0·15 ml kg^{-1} i/m	Immobilization/analgesia
Fentanyl/fluanisone (Hypnorm)	0·1–0·2 ml kg^{-1} i/m	Moderate to heavy sedation, moderate analgesia
Medetomidine	0·1–0·8 mg kg^{-1} i/m, s/c or i/v	Light to heavy sedation, some analgesia
Xylazine	1–2 mg kg^{-1} i/m	Light to moderate sedation, some analgesia

3. Acepromazine (0.2 mg kg^{-1} i/m) alone produces sedation, but has no analgesic action.

4. Fentanyl/fluanisone (Hypnorm, Janssen) (0.1–0.2 ml kg^{-1}) or fentanyl/droperidol (Innovar Vet, Janssen) (0.1–0.15 ml kg^{-1} i/m) produces good analgesia, sufficient for minor surgical procedures and heavy sedation. A moderate bradycardia is often produced, but this can be prevented by administration of atropine (see below). Partial reversal of these agents is possible using nalbuphine or other mixed agonist/antagonist opioids (see Table 7.3).

5. Etorphine/methotrimeprazine (Immobilon SA, C-Vet) (0.5 ml per 4 kg i/m) produces intense analgesia and sedation, which is often sufficient to enable many minor or moderate surgical procedures to be carried out. Bradycardia, hypotension and respiratory depression can be pronounced, although prior administration of atropine can minimize these side-effects. The drug combination can be reversed by administration of diprenorphine (Revivon, C-Vet) (0.5 ml per 4 kg i/v).

6. Xylazine (1–2 mg kg^{-1} i/m) produces good sedation and mild analgesia. Vomiting often occurs after administration and animals may be easily roused by loud noises. Other side-effects include production of bradycardia, occasional heart block and hyperglycaemia; the cardiac effects can be prevented by pre-treatment with atropine.

Atropine (0.05 mg kg^{-1} s/c) or glycopyrrolate (0.01 mg kg^{-1}) should be administered prior to the use of fentanyl/fluanisone, fentanyl/droperidol or xylazine. It may also be included as premedication prior to use of the anaesthetic regimens described below.

The dose rates of anaesthetic drugs listed below and in Table 7.17 can be reduced by 30–50% if one of the drugs (1–6) listed above has been administered.

B. General anaesthesia

1. Injectable agents

Intravenous administration is easily carried out using the cephalic vein on the anterior surface of the forelimb, provided that adequate restraint can be provided.

1. Propofol (5–7.5 mg kg^{-1} i/v) produces a short period (5–10 minutes) of general anaesthesia, which can be prolonged by administration of incremental injections of 1–2 mg kg^{-1} every 10–15 minutes. To

TABLE 7·17
Anaesthetic dose rates in the dog

Drug	Dose rate	Effect	Duration of anaesthesia (min)	Sleep time (min)
Alpha-chloralose	80 mg kg^{-1} i/v	Light anaesthesia	360–600	Non-recovery only
Ketamine/medetomidine	2·5–7·5 mg kg^{-1} i/m + 40 μg kg^{-1} i/m	Light to medium anaesthesia	30–45	60–120
Ketamine/xylazine	5 mg kg^{-1} i/v + 1–2 mg kg^{-1} i/v or i/m	Light to medium anaesthesia	30–60	60–120
Methohexitone	4–8 mg kg^{-1} i/v	Surgical anaesthesia	4–5	10–20
Pentobarbitone	20–30 mg kg^{-1} i/v	Surgical anaesthesia	30–40	60–240
Propofol	5–7·5 mg kg^{-1} i/v	Surgical anaesthesia	5–10	15–30
Thiamylal	10–15 mg kg^{-1} i/v	Surgical anaesthesia	5–10	15–20
Thiopentone	10–20 mg kg^{-1} i/v	Surgical anaesthesia	5–10	20–30
Urethane	1000 mg kg^{-1} i/v	Surgical anaesthesia	360–480	Non-recovery only

Duration of anaesthesia and sleep time (loss of righting reflex) are provided only as a general guide, since considerable between-animal variation occurs. For recommended techniques, see text.

induce anaesthesia, approximately half the computed dose of the drug should be administered rapidly and the remainder injected slowly to effect. Recovery is smooth and rapid, even after prolonged anaesthesia (Glen and Hunter, 1984; Hall and Chambers, 1987; Watkins *et al.*, 1987; Weaver and Raptopoulos, 1990).

2. The barbiturates thiopentone (10–20 mg kg^{-1} of a 1.25% or 2.5% solution), methohexitone (4–8 mg kg^{-1} of a 1% solution) or thiamylal (10–15 mg kg^{-1} of a 2% or 4% solution) can all be used to produce anaesthesia lasting 10–30 minutes, with recovery occurring within 15–20 minutes. Recovery can be associated with involuntary excitement and agitation unless a sedative or tranquillizer has been administered as pre-anaesthetic medication.

3. Pentobarbitone is still widely used in research facilities because it can provide 45–60 minutes of light anaesthesia following a single intravenous dose of the drug (20–30 mg kg^{-1} i/v). It has the disadvantage of providing poor analgesia for major surgical procedures, unless high dosages are administered. At these higher doses, pentobarbitone produces significant respiratory and cardiovascular system depression and it is preferable to ventilate the animal to maintain normal blood gas concentrations.

4. Ketamine is less widely used in the dog than other species, primarily because it may cause behavioural disturbances. Despite these problems, ketamine (2.5–7.5 mg kg^{-1} i/m) and medetomidine (40 μg kg^{-1} i/m) can be used to produce surgical anaesthesia, as can ketamine (5 mg kg^{-1} i/v) and xylazine (1–2 mg kg^{-1} i/v or i/m).

2. *Inhalational agents*

Anaesthesia can be induced using a face-mask in a co-operative animal, or following sedation with one of the drugs listed above in more apprehensive individuals. Even after sedation, some animals may resent this procedure. It is generally preferable to induce anaesthesia with an injectable anaesthetic agent (see above) followed by intubation and maintenance with a volatile anaesthetic. Intubation is a relatively straightforward procedure in the dog. The mouth can be opened wide to provide a clear view of the larynx, so that an endotracheal tube can be passed under direct vision. If difficulty is experienced in visualizing the larynx, a laryngoscope with a Macintosh or Soper blade can be used. Following intubation, the dog can be connected to an appropriate circuit, such as a Bain or Macintosh Magill circuit (see Chapter 3). Halothane (1–2%), methoxyflurane (1.0–1.5%), isoflurane (2–3%) or enflurane (1–2%) all provide stable anaesthesia with good analgesia and muscle relaxation.

V. FERRETS

A. Pre-anaesthetic medication

If a ferret has become accustomed to being handled it can be easily restrained to enable intraperitoneal, intramuscular or intravenous injection of an anaesthetic agent. Some laboratory-reared animals may resent physical restraint, and administration of drugs to sedate the animal may be required before induction of anaesthesia with intravenous or inhalational anaesthetic agents. Ferrets should be starved for 12 hours prior to induction of anaesthesia to minimize the risk of vomiting during induction of anaesthesia.

The following drugs can be used to produce sedation (Table 7.18).

1. Diazepam (2 mg kg^{-1} i/m) produces heavy sedation.
2. Acepromazine (0.2 mg kg^{-1} i/m) produces moderate sedation.
3. Ketamine (20–30 mg kg^{-1} i/m) produces heavy sedation and immobilization lasting about 30–40 minutes.
4. Fentanyl/fluanisone (Hypnorm, Janssen) (0.5 ml kg^{-1} i/m) or fentanyl/droperidol (Innovar Vet) (0.15 ml kg^{-1} i/m) produces heavy sedation and sufficient analgesia for minor surgical procedures.
5. Medetomidine (0.1–0.5 mg kg^{-1} s/c) or xylazine (1–2 mg kg^{-1} i/m) produces light to heavy sedation in the ferret. Sedation can be reversed using atipamezole (1 mg kg^{-1}).

Atropine (0.05 mg kg^{-1} i/m) may be administered if necessary.

TABLE 7.18
Sedatives, tranquillizers and other pre-anaesthetic medication for use in the ferret

Drug	Dose rate	Comments
Acepromazine	0·2 mg kg^{-1} i/m	Moderate sedation
Atropine	0·05 mg kg^{-1} s/c or i/m	Anticholinergic
Diazepam	2 mg kg^{-1} i/m	Light sedation
Fentanyl/fluanisone (Hypnorm)	0·5 ml kg^{-1} i/m	Immobilization, good analgesia
Ketamine	20–30 mg kg^{-1}	Immobilization, some analgesia

B. General anaesthesia

1. Injectable agents

The cephalic vein on the anterior aspect of the foreleg can be used for intravenous injection, although firm restraint is necessary. Intramuscular injections can easily be made into the hindlimb muscles. Anaesthetic dose rates are summarized in Table 7.19.

1. Ketamine (25 mg kg^{-1} i/m) and xylazine (1–2 mg kg^{-1} i/m), or ketamine (4–8 mg kg^{-1} i/m) and medetomidine (0.05–0.1 mg kg^{-1} i/m) (Wolfensohn and Lloyd, 1994) produce good surgical anaesthesia lasting 30–60 minutes. As in other species, administration of atipamezole will speed recovery. Ketamine combined with diazepam (2 mg kg^{-1} i/m) or acepromazine (0.25 mg kg^{-1} i/m) has similar effects to ketamine/xylazine.
2. Alphaxalone/alphadolone (8–12 mg kg^{-1} i/v) produces good surgical anaesthesia lasting 10–15 minutes. Additional doses of the drug can be administered to prolong the period of anaesthesia.
3. Pentobarbitone (25–30 mg kg^{-1} i/v, 36 mg kg^{-1} i/p) can be used to provide 30–120 minutes of light to medium surgical anaesthesia.

2. Inhalational agents

Halothane, methoxyflurane, enflurane and isoflurane can all be used to produce or maintain surgical anaesthesia in ferrets. Animals can be induced in an anaesthetic chamber, following which anaesthesia can be maintained using a face-mask. Alternatively, the animal may be intubated using a paediatric laryngoscope blade.

VI. PIGS

A. Pre-anaesthetic medication

Small pigs (< 10 kg) are easily restrained humanely, but pigs of all sizes vocalize extremely loudly when restrained and it may be thought useful to administer a sedative before induction of anaesthesia. Although it is possible to physically restrain larger pigs, use of pre-anaesthetic medication will considerably ease induction and reduce stress to the animal. Several of the pre-anaesthetic agents require administration of large volumes of drug, and this appears to cause less distress if carried out slowly. A useful technique is to attach a 50 mm needle (or longer in

TABLE 7·19
Anaesthetic dose rates in the ferret

Drug	Dose rate	Effect	Duration of anaesthesia (min)	Sleep time (min)
Alphaxalone/alphadolone	8–12 mg kg^{-1} i/v	Surgical anaesthesia	10–15	20–30
	12–15 mg kg^{-1} i/m	Light anaesthesia	15–30	60–90
Ketamine/acepromazine	25 mg kg^{-1} i/m + 0·25 mg kg^{-1} i/m	Surgical anaesthesia	20–30	60–120
Ketamine/diazepam	25 mg kg^{-1} i/m + 2 mg kg^{-1} i/m	Surgical anaesthesia	20–30	60–120
Ketamine/medetomidine	8 mg kg^{-1} i/m + 0·1 mg kg^{-1} i/m	Surgical anaesthesia	20–30	60–120
Ketamine/xylazine	25 mg kg^{-1} i/m + 1–2 mg kg^{-1} i/m	Surgical anaesthesia	20–30	60–120
Pentobarbitone	25–30 mg kg^{-1} i/v, 36 mg kg^{-1} i/p	Surgical anaesthesia	30–60	90–240
Urethane	1500 mg kg^{-1} i/v	Surgical anaesthesia	360–480	Non-recovery only

Duration of anaesthesia and sleep time (loss of righting reflex) are provided only as a general guide, since considerable between-animal variation occurs. For recommended techniques, see text.

animals over 50 kg) to the syringe using an anaesthetic extension tube. The needle can be quickly placed in the pig's neck muscles, and the injection made slowly with the pig unrestrained.

Pigs are usually starved for 12 hours prior to induction of anaesthesia, although vomiting on induction is rare.

The following drugs can be used for pre-anaesthetic medication (Table 7.20).

1. Diazepam (1–2 mg kg^{-1} i/m) provides rapid sedation but is best followed by administration of ketamine (10–15 mg kg^{-1} i/m) to provide complete immobilization. Some preparations of diazepam can be mixed in the same syringe with ketamine; alternatively midazolam (1–2 mg kg^{-1}) can be used. The large volume of injectate (6–8 ml) required for animals weighing over 15–20 kg limits the use of this combination to smaller pigs.

2. Ketamine (10 mg kg^{-1} i/m) used alone in juvenile and adult pigs immobilizes the animal but spontaneous movements occur. In young animals this drug appears less effective and higher dose rates (20 mg kg^{-1}) may be required. Even at higher doses considerable spontaneous movements may occur. Although this drug is useful in older animals, it is an expensive means of producing sedation.

3. Tiletamine and zolezepam (Zoletil, Virbac) (20 mg kg^{-1}) produces heavy sedation and immobilization. An advantage of this combination is that the volume for injection can be reduced, making it particularly suitable for larger animals.

TABLE 7.20
Sedatives, tranquillizers and other pre-anaesthetic medication for use in the pig

Drug	Dose rate	Comments
Acepromazine	0·2 mg kg^{-1} i/m	Moderate sedation
Alphaxalone/alphadolone	6 mg kg^{-1} i/m	Sedation
Atropine	0·05 mg kg^{-1} s/c or i/m	Anticholinergic
Azaperone	5 mg kg^{-1} i/m	Moderate to deep sedation
Diazepam	1–2 mg kg^{-1} i/m	Light to moderate sedation
Ketamine	10–15 mg kg^{-1} i/m	Sedation, immobilization
Metomidate	2 mg kg^{-1} i/m	Moderate to deep sedation

4. Alphaxalone/alphadolone (6 mg kg^{-1} i/m) produces good sedation in pigs, but the volume of injection (10 ml for a 20 kg pig) limits the use of this drug to smaller animals.

5. Azaperone (5 mg kg^{-1} i/m) produces sedation but has no analgesic effect. It is best combined with the administration of metomidate (2 mg kg^{-1} i/m) to produce deep sedation and sufficient analgesia for minor surgical procedures.

6. Acepromazine (0.2 mg kg^{-1} i/m) produces moderate sedation but has no analgesic action.

7. Medetomidine and xylazine are relatively ineffective in many strains of pigs, and the degree of sedation produced is highly variable, although these agents may potentiate the effects of other anaesthetics.

Atropine (0.05 mg kg^{-1} i/m) can be administered to reduce salivary and bronchial secretions.

B. General anaesthesia

1. Injectable agents

Following physical or chemical restraint, a number of different anaesthetic agents can be administered by intravenous injection to produce surgical anaesthesia. The dose rates of anaesthetic drugs listed below and in Table 7.21 can be reduced by 30–50% if one of the drugs (1–6) listed above has been administered. The most convenient route is via the ear veins, and placement of an over-the-needle catheter to ensure reliable venous access is strongly recommended.

1. Propofol (2.5–3.5 mg kg^{-1} i/v) produces surgical anaesthesia lasting 10 minutes. As with other species a short period of apnoea often occurs immediately after injection. Anaesthesia can be prolonged by administration of incremental injections (1–2 mg kg^{-1} every 10–15 minutes) or by continuous infusion (8–9 mg kg^{-1} h^{-1}). If propofol is used as the sole anaesthetic agent for major surgical procedures, significant respiratory depression may occur, and ventilation may need to be assisted. An alternative approach is to administer propofol at a lower rate (5–6 mg kg^{-1} h^{-1}) together with alfentanil (20–30 μg kg^{-1} i/v followed by 2–5 μg kg^{-1} min^{-1}) to provide supplemental analgesia, although use of alfentanil will require assisted ventilation. Recovery following propofol is smooth and rapid, as with other species, and if necessary the alfentanil can be reversed by administration of nalbuphine (0.5 mg kg^{-1} i/v).

TABLE 7.21
Anaesthetic dose rates in the pig

Drug	Dose rate	Effect	Duration of anaesthesia (min)	Sleep time (min)
Alphaxalone/alphadolone	6 mg kg^{-1} i/m then 2 mg kg^{-1} i/v	Immobilization, Surgical anaesthesia	– 5–10	10–20 15–20
Azaperone + metomidate	5 mg kg^{-1} i/m + 3·3 mg kg^{-1} i/v	Light to medium anaesthesia	30–40	60–90
Ketamine	10–15 mg kg^{-1} i/m	Sedation, immobilization	20–30	60–120
Ketamine/acepromazine	22 mg kg^{-1} + 1·1 mg kg^{-1} i/m	Light anaesthesia	20–30	60–120
Ketamine/diazepam	10–15 mg kg^{-1} i/m + 0·5–2 mg kg^{-1} i/m	Immobilization/ Light anaesthesia	20–30	60–90
Ketamine/medetomidine	10 mg kg^{-1} i/m + 0·08 mg kg^{-1} i/m	Immobilization/Light anaesthesia	40–90	120–240
Ketamine/midazolam	10–15 mg kg^{-1} i/m + 0·5–2 mg kg^{-1} i/m	Immobilization/Light anaesthesia	20–30	60–90
Methohexitone	5 mg kg^{-1} i/v	Surgical anaesthesia	4–5	5–10
Pentobarbitone	20–30 mg kg^{-1} i/v	Light to surgical anaesthesia	20–30	60–120
Propofol	2·5–3·5 mg kg^{-1} i/v	Surgical anaesthesia	5–10	10–20
Thiopentone	6–9 mg kg^{-1} i/v	Surgical anaesthesia	5–10	10–20

Drug	Dose rate	Effect	Duration of anaesthesia (min)	Sleep time (min)
Tiletamine/zolezepam	2–4 mg kg^{-1} i/m	Immobilization	20–30	60–120
	6–8 mg kg^{-1} i/m	Light anaesthesia	20–30	90–180
Tiletamine/zolezepam + xylazine	2–7 mg kg^{-1} i/m + 0·2–1 mg kg^{-1} i/m	Light to medium anaesthesia	30–40	60–120

Duration of anaesthesia and sleep time (loss of righting reflex) are provided only as a general guide, since considerable between-animal variation occurs. For recommended techniques, see text.

2. Alphaxalone/alphadolone (1–2 mg kg^{-1} i/v, to effect) will produce surgical anaesthesia and good muscle relaxation with minimal respiratory depression. Prolonged anaesthesia can be achieved by continuous infusion of this drug.

3. Methohexitone (5 mg kg^{-1} i/v) or thiopentone (6–9 mg kg^{-1} i/v) can be administered to produce 5–10 minutes of surgical anaesthesia. Recovery can be associated with excitement in some individuals unless a sedative or tranquillizer has been administered.

4. Pentobarbitone (20–30 mg kg^{-1} i/v) produces light surgical anaesthesia in pigs. The high dose rates (30 mg kg^{-1}) needed to produce deep surgical anaesthesia may cause severe cardiovascular system depression. The duration of anaesthesia in pigs is shorter than in other species with surgical anaesthesia persisting for only 20–30 minutes; however, full recovery can take 3–4 hours.

2. *Inhalational agents*

It is possible to produce anaesthesia with volatile anaesthetics administered by means of a face-mask, but providing the necessary degree of physical restraint can be a problem, even in small pigs. Considerable pollution of the environment with anaesthetic gases will occur and this should be avoided if possible. It is preferable to induce anaesthesia with an injectable anaesthetic drug, intubate the pig and maintain anaesthesia using an inhalational agent. Halothane, methoxyflurane, enflurane and isoflurane can all be used to maintain safe and effective anaesthesia. The potency of volatile anaesthetics appears lower in pigs than in other laboratory mammals, so slightly higher concentrations are required for induction and maintenance. A small number of pigs have been shown to develop malignant hyperthermia in response to halothane anaesthesia and, whenever possible, animals should be obtained from herds that have a low incidence of this problem.

Pigs should be maintained using a Bain or Magill circuit, but large animals (> 30 kg) may require closed-circuit anaesthesia in order to reduce the gas flow rates needed. As discussed in Chapter 3, closed-circuit anaesthesia requires considerable expertise, and advice should be sought from a veterinary surgeon or medically qualified anaesthetist if this type of circuit is to be used.

Intubation of pigs is complicated by the difficulty of obtaining a clear view of the larynx and by the laryngeal anatomy which tends to obstruct passage of the tube. A brief description of the technique of intubation is given in Chapter 3. More detailed descriptions are given by Hall and Clarke (1991).

VII. SHEEP AND GOATS

A. Pre-anaesthetic medication

Opinion varies as to whether sheep and goats should be starved before induction of general anaesthesia. Fasting has little effect on the volume of digesta present in the rumen, but may reduce the incidence of ruminal tympany (an accumulation of gas in the rumen caused by bacterial fermentation). This appears to be a greater problem in animals that are grazing. Unnecessary fasting should be avoided since it may cause distress to the animal, and the author has experienced few problems if food and water are provided up to an hour prior to induction of anaesthesia. If ruminal tympany develops it can be relieved by passage of a stomach tube (see Chapter 4). Should the condition occur repeatedly, pre-anaesthetic fasting may be introduced, although this may not resolve the problem.

Sheep and goats can generally be restrained easily for administration of anaesthetic agents. The stress associated with movement from its pen to the operating theatre can be reduced by use of sedatives and tranquillizers. High doses of sedatives and tranquillizers can also be used in conjunction with local anaesthetic agents to provide humane restraint and surgical anaesthesia. A useful review of anaesthetic techniques in sheep and goats is provided by Taylor (1991).

The following drugs may be used for pre-anaesthetic medication (Table 7.22).

1. Diazepam (2 mg kg^{-1} i/m or 1 mg kg^{-1} i/v) is a particularly effective tranquillizer in sheep and goats, as is midazolam (0.5 mg kg^{-1} i/v). When combined with ketamine (4 mg kg^{-1} i/v), light to moderate surgical anaesthesia is produced.

2. Xylazine (0.2 mg kg^{-1} i/m) will provide heavy sedation and good analgesia in sheep, lasting 30–35 minutes. It can be combined with ketamine (4 mg kg^{-1} i/v) to produce light surgical anaesthesia. Goats appear more sensitive to xylazine and lower doses of 0.05 mg kg^{-1} i/m are usually adequate.

3. Medetomidine (25 μg kg^{-1} i/m) is also effective in producing sedation and analgesia. As with xylazine, this agent can be combined with ketamine to produce surgical anaesthesia, but only a low dose of ketamine (1 mg

TABLE 7.22
Sedatives, tranquillizers and other pre-anaesthetic medication for use in the sheep and goat

Drug	Dose rate	Comments
Acepromazine	0·05–0·1 mg kg^{-1} i/m	Moderate sedation
Diazepam	2 mg kg^{-1} i/m, 1 mg kg^{-1} i/v	Light to moderate sedation
Ketamine	20 mg kg^{-1} i/m	Moderate to heavy sedation, immobilization, some analgesia
Medetomidine	25 μg kg^{-1} i/m	Light to heavy sedation, some analgesia
Midazolam	0·5 mg kg^{-1} i/v	Moderate sedation
Xylazine	0·2 mg kg^{-1} i/m (sheep), 0·05 mg kg^{-1} i/m (goat)	Light to moderate sedation, some analgesia

kg^{-1} i/m) is required because of the potency of medetomidine in this species (Laitinen, 1990). Both medetomidine and xylazine can produce hypoxia in sheep owing to pulmonary effects (Celly *et al.*, 1994; Tulamo *et al.*, 1995). Sedation, and any undesirable side-effects, can be reversed using atipamezole (100–200 μg kg^{-1} i/v or i/m).

4. Acepromazine (0.05–0.1 mg kg^{-1} i/m) will sedate sheep but provides no analgesia.

The use of atropine in sheep is of limited value. Extremely high dose rates (0.5 mg kg^{-1} i/m) are needed to reduce salivary secretions, and repeated doses of 0.2–0.3 mg kg^{-1} may be required every 15 minutes to maintain this drug's effects.

B. General anaesthesia

1. Injectable agents

Intravenous injection is easily carried out using the marginal ear vein, anterior cephalic vein on the foreleg, or jugular vein. Dose rates of injectable anaesthetic drugs are summarized in Table 7.23.

1. Light to moderate surgical anaesthesia can be produced by use of

TABLE 7.23
Anaesthetic dose rates in the sheep and goat

Drug	Dose rate	Effect	Duration of anaesthesia (min)	Sleep time (min)
Alphaxalone/ alphadolone	2–3 mg kg^{-1} i/v (adult), 6 mg kg^{-1} i/v (lamb or kid)	Surgical anaesthesia	5–10	10–20
Etorphine/acepromazine (Immobilon LA)	0·5 ml per 50 kg i/m (> 30 kg)	Immobilization, analgesia	Analgesia 30–40	60–90
Etorphine/ methotrimeprazine (Immobilon SA)	0·5 ml per 4 kg i/m (< 30 kg)	Immobilization, analgesia	Analgesia 30–40	60–90
Ketamine/diazepam	10–15 mg kg^{-1} + 2 mg kg^{-1} i/m, or 4 mg kg^{-1} i/v + 1 mg kg^{-1} i/v	Light to medium anaesthesia surgical anaesthesia	20–30 20–30	60–90 45–90
Ketamine/ medetomidine	1 mg kg^{-1} i/m + 25 μg kg^{-1} i/m	Surgical anaesthesia	30–60	60–90
Ketamine/xylazine	4 mg kg^{-1} + 0·2 mg kg^{-1} i/v (sheep), 0·05 mg kg^{-1} i/v (goat)	Surgical anaesthesia	15–20	30–90
Methohexitone	4 mg kg^{-1} i/v	Surgical anaesthesia	4–5	5–10
Pentobarbitone	30 mg kg^{-1} i/v	Immobilization, anaesthesia	15–30	30–60
Propofol	4–5 mg kg^{-1} i/v	Light anaesthesia	5–10	10–15
Thiopentone	10–15 mg kg^{-1} i/v	Surgical anaesthesia	5–10	10–20
Urethane	1000 mg kg^{-1} i/v	Surgical anaesthesia	360–480	Non-recovery only

Duration of anaesthesia and sleep time (loss of righting reflex) are provided only as a general guide, since considerable between-animal variation occurs. For recommended techniques, see text.

ketamine (4 mg kg^{-1} i/v) plus xylazine (1 mg kg^{-1} i/v), or ketamine (10–15 mg kg^{-1} i/m or 4 mg kg^{-1} i/v) plus diazepam or midazolam (2 mg kg^{-1} i/m or 1 mg kg^{-1} i/v) (Coulson *et al.*, 1991), as described above. The combination of ketamine and diazepam appears to cause less cardiovascular and respiratory system depression than does ketamine with xylazine.

2. Thiopentone (10–15 mg kg^{-1} i/v) and methohexitone (4 mg kg^{-1} i/v) both provide 5–10 minutes of anaesthesia.

3. Alphaxalone/alphadolone (2–3 mg kg^{-1} i/v in adults, 6 mg kg^{-1} i/v in lambs) provides excellent stable anaesthesia in sheep and lambs (Eales and Small, 1982) and can be administered by continuous infusion to maintain anaesthesia over prolonged periods.

4. Propofol (4–5 mg kg^{-1} i/v) can be used to induce anaesthesia in sheep and goats (Waterman, 1988).

5. Pentobarbitone (30 mg kg^{-1} i/v) produces anaesthesia lasting 15–30 minutes. The dose required varies considerably in different animals and the drug generally produces marked respiratory depression.

2. *Inhalational agents*

Sheep may be restrained and anaesthesia induced using a face-mask, but considerable pollution of the environment with inhalational anaesthetics will occur because of the high fresh gas flows used when anaesthetizing these larger animals. It is generally preferable to induce anaesthesia with an injectable agent, intubate the animal and maintain anaesthesia with an inhalational agent if required.

It is essential to intubate sheep immediately following induction of general anaesthesia since regurgitation of rumen contents invariably occurs and these may be inhaled, resulting in an inhalational pneumonia. Intubation is relatively straightforward, provided that a suitable laryngoscope blade (see Table 3.1) is available. The vocal cords should be sprayed with lignocaine prior to passage of an endotracheal tube to prevent laryngospasm.

Halothane, methoxyflurane, enflurane and isoflurane can all be used to maintain effective anaesthesia in sheep. Animals should be maintained on a Bain or Magill circuit. It may be advisable to use a circle system in large adult sheep to reduce the quantities of anaesthetic gases needed to prevent rebreathing with these circuits.

C. Anaesthetic management

Following intubation it is advisable to pass a stomach tube to try to minimize the risk of ruminal tympany developing. Sheep should, if possible, be positioned on their sides; in dorsal recumbency, the pressure of the rumen and other viscera on the major abdominal blood vessels may interfere with venous return. During the post-operative recovery period, sheep should be positioned on their sternum and observed carefully for signs of ruminal tympany.

VIII. PRIMATES

A. Pre-anaesthetic medication

Small primates such as marmosets (*Callithrix jacchus*) can usually be restrained easily enough to enable the intramuscular injection of anaesthetic agents. Pre-anaesthetic medication to sedate these animals is generally only required prior to intravenous injection of induction agents, or to enable administration of inhalational agents by means of a face mask. Larger primates such as baboons (*Papio* sp.) and macaque monkeys (*Macaca* sp.) can cause physical injury to their handler if inexpertly restrained, and it is strongly recommended that chemical agents be used to produce deep sedation.

The following drugs may be used for pre-anaesthetic medication (Table 7.24).

1. Ketamine (5–25 mg kg^{-1} i/m) is probably the drug of choice in larger primates. At the lower dose rate, heavy sedation is produced; higher doses of ketamine produce light surgical anaesthesia (Banknieder *et al.*, 1978; White and Cunnings, 1979).
2. Alphaxalone/alphadolone (12–18 mg kg^{-1} i/m) is the agent of choice for sedating marmosets and small primates. Heavy sedation is produced and additional doses of the drug can be administered intravenously to produce surgical anaesthesia (Phillips and Grist, 1975).
3. Acepromazine (0.2 mg kg^{-1} i/m) produces sedation but will not immobilize the animal.
4. Diazepam (1 mg kg^{-1} i/m) sedates primates, but provides insufficient sedation for the safe handling of large primates.
5. Fentanyl/fluanisone (Hypnorm, Janssen) (0.3 ml kg^{-1} i/m) or fentanyl/droperidol (Innovar Vet) (0.3 ml kg^{-1} i/m) produces heavy sedation and good analgesia (Field *et al.*, 1966).

TABLE 7.24
Sedatives, tranquillizers and other pre-anaesthetic medication for use in the non-human primate

Drug	Dose rate	Comments
Acepromazine	0·2 mg kg^{-1} i/m	Moderate sedation
Alphaxalone/alphadolone	12–18 mg kg^{-1} i/m	Heavy sedation
Atropine	0·05 mg kg^{-1} s/c or i/m	Anticholinergic
Diazepam	1 mg kg^{-1} i/m	Light to moderate sedation
Fentanyl/droperidol (Innovar)	0·3 ml kg^{-1} i/m	Heavy sedation, good analgesia
Fentanyl/fluanisone (Hypnorm)	0·3 ml kg^{-1} i/m	Heavy sedation, good analgesia
Ketamine	5–25 mg kg^{-1} i/m	Moderate sedation, immobilization, some analgesia
Xylazine	0·5 mg kg^{-1} i/m	Light to moderate sedation, some analgesia

Atropine (0.05 mg kg^{-1} i/m) should be administered to minimize the bradycardia produced by neuroleptanalgesic combinations and to reduce the amount of salivary secretions.

Primates should be starved for 12–16 hours prior to induction of anaesthesia.

B. General anaesthesia

1. Injectable agents

The cephalic vein on the anterior aspect of the forelimb of large primates, or the lateral tail vein in marmosets can be used for intravenous injection. Dose rates of injectable anaesthetics are summarized in Table 7.25.

1. The combination of ketamine (10 mg kg^{-1} i/m) and xylazine (0.5 mg kg^{-1} i/m) produces surgical anaesthesia with good muscle relaxation, lasting 30–40 minutes. Ketamine (15 mg kg^{-1} i/m) plus diazepam (1 mg kg^{-1} i/m) has similar effects.
2. Alphaxalone/alphadolone (10–12 mg kg^{-1} i/v) produces good surgical

TABLE 7.25
Anaesthetic dose rates in the non-human primate

Drug	Dose rate	Effect	Duration of anaesthesia (min)	Sleep time (min)
Alphaxalone/alphadolone	10–12 mg kg^{-1} i/v	Surgical anaesthesia	5–10	10–20
	12–18 mg kg^{-1} i/m	Immobilization, anaesthesia	10–20	30–50
Ketamine/diazepam	15 mg kg^{-1} i/m + 1 mg kg^{-1} i/m	Surgical anaesthesia	30–40	60–90
Ketamine/xylazine	10 mg kg^{-1} i/m + 0·5 mg kg^{-1} i/m	Surgical anaesthesia	30–40	60–120
Methohexitone	10 mg kg^{-1} i/v	Surgical anaesthesia	4–5	5–10
Pentobarbitone	25–35 mg kg^{-1} i/v	Surgical anaesthesia	30–60	60–120
Propofol	7·5–12·5 mg kg^{-1} i/v	Surgical anaesthesia	5–10	10–15
Thiopentone	15–20 mg kg^{-1} i/v	Surgical anaesthesia	5–10	10–15

Duration of anaesthesia and sleep time (loss of righting reflex) are provided only as a general guide, since considerable between-animals and between-species variation occurs. For recommended techniques, see text.

anaesthesia in both Old and New World primates, and prolonged periods of anaesthesia can be provided by administration of additional doses (5 mg kg^{-1} i/v) every 10–15 minutes or by continuous infusion (Cookson and Mills, 1983).

3. Propofol (7.5–12.5 mg kg^{-1} i/v) can be used to induce and maintain anaesthesia in both marmosets and larger primates. Good surgical anaesthesia is produced with rapid and smooth recoveries (Glen, 1980; Sainsbury *et al.*, 1991).

4. Methohexitone (10 mg kg i/v) or thiopentone (15–20 mg kg^{-1} i/v) can be administered to produce 5–10 minutes of surgical anaesthesia. The dosage of these agents can be reduced by at least 50% if the animal has received ketamine as pre-anaesthetic medication.

5. Pentobarbitone (25–35 mg kg^{-1} i/v) will provide 30–60 minutes of light surgical anaesthesia, but severe respiratory depression often occurs at higher dose rates. The dose of pentobarbitone should be reduced by at least 50% if ketamine or other sedatives have been administered as pre-anaesthetic medication.

2. *Inhalational agents*

All the commonly available inhalational anaesthetics can be administered to primates. It is usually most convenient to induce anaesthesia using injectable agents. Small primates can then be maintained using an inhalational agent delivered by a face mask, but with larger primates it is preferable to intubate the animal. Endotracheal intubation is relatively straightforward in larger primates using a Magill laryngoscope blade. The larynx should be sprayed with lignocaine before attempting to pass an endotracheal tube. Following intubation, a T-piece or Bain circuit should be used for administration of oxygen and the inhalational anaesthetic.

IX. OTHER SPECIES

A full description of all of the available anaesthetic techniques for fish, reptiles, amphibia and birds is outside the scope of this book, but the following section gives initial guidance on anaesthesia of these species. Further information is available in a number of excellent reviews and textbooks (Green, 1981b; Samour *et al.*, 1984; Harrison and Harrison, 1986; Schaeffer *et al.*, 1992; Brown, 1993).

A. Birds

A number of unique aspects of avian physiology influence the selection of anaesthetic agents and the overall management of anaesthesia. Birds, particularly smaller species, have higher metabolic rates than mammals of comparable size and have higher body temperatures. The higher body temperature results in an increased temperature gradient between core temperature and the external environmental temperature, and so cooling during anaesthesia is rapid. In small birds, as with mammals, the high ratio of surface area to body weight also increases heat loss. It is therefore particularly important to adopt measures to minimize heat loss (Green, 1981b). In addition to the precautions described in Chapter 4, avoid removing large numbers of feathers. Because of their high metabolic rate, small birds do not tolerate fasting and may develop hypoglycaemia, so only individuals weighing more than 1 kg should undergo pre-anaesthetic fasting.

The respiratory system in birds differs from that of mammals, having a series of air sacs which connect with the lungs. The presence of air sacs may allow a build-up of anaesthetic vapour in dependent areas of the respiratory system, with consequent overdosage. In any event, induction of anaesthesia with volatile agents is rapid in birds. This is not a problem provided that the anaesthetist appreciates the short time in which induction may occur.

When positioning the bird for surgery, avoid taping the wings and legs in full extension as this can inhibit both respiratory movements and venous return. Birds should be handled gently at all times, taking particular care to avoid obstructing respiratory movements, since even short periods of apnoea can result in hypoxia. As with small mammals, care must be taken to avoid laying instruments across the chest or resting the operator's hands on the bird, since this can easily impede respiration in small species.

In the post-anaesthetic period, continued attention to maintaining body temperature is essential. Birds should recover in a heated incubator or cage (40°C for small birds, 35°C for species over 250 g). The recovery area should be quiet, with subdued lighting. Prolonged recovery can be associated with episodes of wing flapping, and if not restrained the bird may injure itself. Placing a temporary bandage or cloth around the wings can help to control this problem. Such difficulties can be largely avoided, however, by using isoflurane for anaesthesia, since recovery from this agent is extremely rapid. Very little information is available concerning the use of analgesics for post-operative pain relief in birds. No clinical trials of their efficacy have been undertaken, and only very limited data are available for analgesiometry. Suggestions based on clinical impression include buprenorphine (0.05 mg kg^{-1} i/m), butorphanol (2–4 mg kg^{-1} i/m), or the NSAIDs flunixin (1–10 mg kg^{-1}) (Harrison and Harrison, 1986) or ketoprofen (2 mg kg^{-1}) (Wolfensohn and Lloyd, 1994). Dose rates of anaesthetics and analgesics are summarized in Table 7.26.

1. Pre-anaesthetic medication

a. Ketamine. Ketamine can be administered to a wide range of avian species, but its effects can vary considerably. In the domestic fowl, 15–20 mg kg^{-1} i/m) produces immobilization and some individuals may be lightly anaesthetized. Following initial chemical restraint with ketamine, anaesthesia can be deepened using isoflurane. When using ketamine in small birds (e.g. small finches, body weight 20–49 g), the required dose (30–40 mg kg^{-1}) is best administered by diluting the commercial veterinary preparation with saline to provide a 10 mg ml^{-1} solution. A volume of 0.1 ml per bird i/m will then provide heavy sedation, whilst 0.2 ml will produce light anaesthesia (Green, 1981b).

b. Metomidate. Metomidate (10–20 mg kg^{-1} i/m, 5% solution) sedates and immobilizes many species of bird. It is particularly effective in domestic fowl. When anaesthetizing small birds (< 100 g), a dilute solution (0.1%) should be used.

2. General anaesthesia

a. Injectable agents.

1. Ketamine, in combination with xylazine, will produce light to moderate surgical anaesthesia in birds. As with ketamine alone, the effects vary between species, and it may be necessary to carry out a pilot study to assess the response of the species to be anaesthetized. Allometric scaling of ketamine/xylazine doses appears to provide a reasonable guide to an appropriate dose rate (Harrison and Harrison, 1986), with a dose range of 10–30 mg kg^{-1} ketamine and 2–6 mg kg^{-1} xylazine, smaller birds requiring higher dose rates per kg. Ketamine (20–40 mg kg^{-1}) and diazepam (1–1.5 mg kg^{-1}) or midazolam (4 mg kg^{-1}), all given by intramuscular injection, produce light to medium surgical anaesthesia in birds, but as with other ketamine combinations, dose rates vary with the body weight and species involved.

2. Alphaxalone/alphadolone has unpredictable effects when given intramuscularly (Green, 1981b), but can be administered by the intravenous

TABLE 7·26
Anaesthetic, sedative and analgesic drugs for use in birds

Drug	Dose rate	Effect	Duration of anaesthesia (min)	Sleep time (min)
Alphaxalone/alphadolone	10–14 mg kg^{-1} i/v	Light anaesthesia	10–15	20–60
Buprenorphine	0·01–0·05 mg kg^{-1} i/m	Analgesia	?	
Butorphanol	2–4 mg kg^{-1} i/m	Analgesia	?	
Equithesin*	2·5 ml kg^{-1} i/m	Light to medium anaesthesia	20–30	60–120
Flunixin	1–10 mg kg^{-1} i/m	Analgesia	?	
Ketamine > 1 kg < 1 kg	15–20 mg kg^{-1} i/m, 30–40 mg kg^{-1} i/m	Immobilization, some analgesia	20–30	30–90
Ketamine/diazepam	20–40 mg kg^{-1} i/m + 1–1·5 mg kg^{-1} i/m	Medium surgical anaesthesia	20–30	30–90
Ketamine/midazolam	20–40 mg kg^{-1} i/m + 4 mg kg^{-1} i/m	Medium surgical anaesthesia	20–30	30–90
Ketamine/xylazine	10–30 mg kg^{-1} i/m + 2–6 mg kg^{-1} i/m	Light to medium surgical anaesthesia	10–30	30–60
Ketoprofen	2 mg kg^{-1} s/c	Analgesia		
Metomidate	10–20 mg kg^{-1} i/m	Immobilization	10–30	30–60

* See Appendix 3.

route (10–14 mg kg^{-1}) in larger species to produce light surgical anaesthesia.

3. Equithesin is a combination of pentobarbitone, chloral hydrate and magnesium sulphate (see Appendix 3). Given at a dosage of 2.5 ml kg^{-1} i/m it produces medium planes of surgical anaesthesia in pigeons and domestic fowl. It provides effective anaesthesia in domestic fowl chicks (0.15 ml per 40 g chick i/p), although occasional post-anaesthetic mortality may occur.

b. Inhalational agents. Isoflurane is widely regarded as the anaesthetic agent of choice for birds. It provides smooth and rapid induction of anaesthesia (4–5%), and anaesthesia may be maintained with concentrations of 2–3%. The depth of anaesthesia can be changed very rapidly by changing the inspired concentration, allowing birds to be maintained at an anaesthetic plane appropriate to the degree of surgical stimulus. This feature probably contributes to this agent's reputation for providing 'safe' anaesthesia, since unnecessarily deep planes of anaesthesia can be avoided. Induction can be achieved using a face-mask or by placing the bird in an anaesthetic chamber. Recovery is rapid, and usually free from involuntary excitement.

Halothane can be used to provide surgical anaesthesia in birds, but the margin of safety appears to be considerably less than that provided by isoflurane. In addition, recovery is more prolonged, and birds may be inappetent (Harrison and Harrison, 1986). Provided that the bird is observed carefully, however, induction with 3–4% halothane and maintenance with 1.5–2% provides reasonably safe and effective anaesthesia.

B. Reptiles

Anaesthesia of snakes can be complicated by their ability to hold their breath for several minutes. This can slow the speed of induction when using volatile anaesthetics, and can cause alarm when breath-holding occurs after administration of injectable agents. Breath-holding is a less common problem in chelonians and lizards, but may still occur. Once anaesthesia has been induced, intubation is relatively simple. Recovery from anaesthesia in these species is particularly influenced by environmental temperature; however, it is not advisable to try to speed recovery by placing the animals in a very warm environment. Animals recover best at temperatures of 22–24 °C (Green, 1981b). Small snakes should fast for 24 hours before anaesthesia, and larger species for 7 days. Chelonians and lizards do not require pre-anaesthetic fasting.

1. General anaesthesia

a. Injectable agents.

1. Ketamine is the most widely used injectable anaesthetic for reptiles, producing light to moderate anaesthesia in most species. In snakes, doses of 50 mg kg^{-1} i/m produce sedation, whilst 50–80 mg kg^{-1} i/m result in light to moderate anaesthesia (Cooper, 1974). The effects of ketamine may persist for 1–2 days. Chelonians are usually lightly anaesthetized at dose rates of 60 mg kg^{-1} i/m (Green, 1981b), although once again recovery can take up to 24 hours. Lizards are generally lightly anaesthetized at dose rates of 25–50 mg kg^{-1} i/m (Cooper, 1984), and recovery may take up to 6 hours. In all species, anaesthesia may be deepened by administration of volatile anaesthetics.

2. Alphaxalone/alphadolone (6–9 mg kg^{-1} i/v) can be used to produce surgical anaesthesia, and is particularly effective in chelonians.

3. Propofol (14 mg kg^{-1} i/v in chelonians, 10 mg kg^{-1} in other reptiles) can also be administered to provide surgical anaesthesia (Meredith A., 1995, personal communication).

b. Inhalational agents. Isoflurane, halothane and methoxyflurane can all be used to produce safe and effective anaesthesia. Use of an anaesthetic chamber is convenient for most smaller species of reptile. Anaesthesia can then be maintained using a face-mask or the animal can be intubated and maintained on an appropriate anaesthetic circuit.

C. Amphibia

Anaesthesia of amphibia can be achieved either by injection of the anaesthetic agent or by inhalation. Alternatively, the animal can be placed in water or a moist environment and liquid anaesthetic or anaesthetic vapour added. Absorption occurs through the skin, resulting in induction of anaesthesia. During anaesthesia and recovery, the skin must be kept moist in frogs and newts, but complete immersion in water must be avoided until the animal has regained consciousness, otherwise it might drown. Pre-anaesthetic fasting is not necessary.

Hypothermia has been suggested as a suitable means of immobilizing amphibia for surgery, but this technique is not considered humane since the degree of analgesia produced is unknown.

1. General anaesthesia

a. Injectable agents. Immersion in tricaine methanesulphonate (MS222) 0.05–0.5% rapidly induces anaesthesia. The animal can then be removed and anaesthesia maintained by wrapping the animal in a cloth soaked in the anaesthetic solution (Green, 1981b). Recovery time can be reduced by washing the animal in water to remove surplus anaesthetic.

b. Inhalational agents. Methoxyflurane can be used to provide safe and effective anaesthesia in amphibia, by exposure to the vapour in an anaesthetic chamber. During induction the base of the chamber should be lined with moist cotton wool to prevent drying of the skin.

D. Fish

Fish are most easily anaesthetized by immersion in anaesthetic solution. Since these animals may be sensitive to sudden changes in pH and temperature, it may be advisable to use some of the water from their normal tank to fill the anaesthetic chamber (Jolly *et al.*, 1972). Following induction of anaesthesia, the fish can be removed from the solution of anaesthetic, wrapped in moist gauze to prevent desiccation, and any procedure undertaken rapidly. For some procedures, it is possible to position the fish so that its gills remain submerged in anaesthetic solution. Alternatively, a more complex system, in which oxygenated anaesthetic solution is passed over the gills, can be constructed. A simpler recirculating system has been described by Brown (1987). It is important to minimize handling of the fish during anaesthesia, since the skin is easily damaged, resulting in post-operative infections. Fish should be fasted for 24–48 hours prior to anaesthesia, as they may vomit and this can interfere with gill function.

The signs of onset of anaesthesia in fish have been described in detail (Green, 1981b), and differ significantly from those in mammals. Briefly, after loss of equilibrium and muscle tone, and onset of very shallow opercular movements, the response to pressure on the muscles at the tail base is reduced, but not abolished. At this stage the fish can be removed from the anaesthetic solution and surgery or other manipulations carried out. If surgical stimuli cause muscle spasms, then the fish can either be returned to the anaesthetic solution, or additional solution can be dripped or sprayed over the gills, for example by placing a drip in the buccal cavity. Overdosage is indicated by loss of regular opercular movements, and occasional exaggerated respiratory movements. Cardiac arrest follows in

1–2 minutes unless the fish is resuscitated. This can be achieved either by flushing the mouth, and hence the gills, with fresh water, or by placing the fish in a tank of fresh water and moving it back and forth with its mouth open.

1. General anaesthesia

a. Injectable agents. Tricaine methanesulphonate (MS222) is administered as a 25–300 mg l^{-1} solution, by immersion; the concentration used determines the degree of anaesthesia. Most small to medium-sized fish (e.g. goldfish, trout) require 100 mg l^{-1} for surgical anaesthesia. Anaesthesia is induced in around 2 minutes, and recovery occurs about 5 minutes after removal from the anaesthetic solution.

Benzocaine should be administered as a freshly prepared solution of 200 mg benzocaine in 5 ml acetone (Green, 1981b), which when added to 8 litres of water provides a solution of 25 p.p.m. (25 mg l^{-1}). This concentration is sedative, enabling minor manipulations to be undertaken. Higher concentrations (50 p.p.m., 50 mg l^{-1}) induce surgical anaesthesia. Some species (e.g. *Tilapia*) require higher concentrations (100 p.p.m., 100 mg l^{-1}) (Iwama, 1992)

References

Aeschbacher, G., and Webb, A.I. (1993a). Propofol in rabbits. 1. Determination of an induction dose. *Laboratory Animal Science* **43**, 324–327.

Aeschbacher, G., and Webb, A.I. (1993b). Propofol in rabbits. 2. Long-term anesthesia. *Laboratory Animal Science* **43**, 328–335.

Alexander, J.I., and Hill, R.G. (1987). "Postoperative Pain Control". Blackwell, Oxford.

Allen, J. (1992). Pulse oximetry: everyday uses in zoological practice. *Veterinary Record* **131**, 354–355.

Allen, D.G., Dyson, D.H., Pascoe, P.J., and O'Grady, M.R. (1986). Evaluation of a xylazine–ketamine hydrochloride combination in the cat. *Canadian Journal of Veterinary Research* **50**, 23–26.

Amrein, R., and Hetzel, W. (1990). Pharmacology of Dormicum (midazolam) and Anexate (flumazenil). *Acta Anaesthesiologica Scandinavica* **34**, suppl. 92, 6–15.

Andrews, C.J.H., and Prys-Roberts, C. (1983). Fentanyl—a review. *In* "Opiate Analgesia" (R.E.S. Bullingham, ed.). pp. 97–122. Saunders, London.

Association of Veterinary Teachers and Research Workers, (1986) Guidelines for the recognition and assessment of pain in animals. *Veterinary Record* **118**, 334–338.

Attia, R.R., Grogono, A.W., and Domer, F.R. (1987). "Practical Anaesthetic Pharmacology", 2nd Ed. Appleton-Century-Crofts, Norwalk, Connecticut.

Ayre, P. (1956). The T-Piece technique. *British Journal of Anaesthesia* **28**. 520–523.

Bain, J., and Spoerel, W.E. (1972). A streamlined anaesthetic system. *Canadian Anaesthesia Society Journal* **19**, 426–435.

Banknieder, A.R., Phillips, J.M., Jackson, K.T., and Vinal, Jr., S.I. (1978). Comparison of ketamine with the combination of ketamine and xylazine for effective anaesthesia in the rhesus monkey. *Laboratory Animal Science* **28**, 742–745.

Barzago, M.M., Bortolotti, A., Stellari, F.F., Pagani, C., Marraro, G., and Bonati, M. (1994). Respiratory and hemodynamic functions, blood-gas parameters, and acid–base balance of ketamine–xylazine anesthetized guinea pigs. *Laboratory Animal Science* **44**, 648–650.

Bendixen, H.H. (1984). The tasks of the anaesthesiologist. In: "Monitoring in Anaesthesia" (L.J. Saidman, and N. Ty Smith, eds). pp. xi–xvi. London, Butterworths.

Beyers, T., Richardson, J.A., and Prince, M.D. (1991). Axonal degeneration and self-mutilation as a complication of the intramuscular use of ketamine and xylazine in rabbits. *Laboratory Animal Science* **41**, 519–520.

Beynen, A.C., Baumans, V., Bertens, A.P.M.G., Havenaar, R., Hesp, A.P.M., and Van Zutphen, L.F.M. (1987). Assessment of discomfort in gallstone-bearing mice: a practical example of the problems encountered in an attempt to recognise discomfort in laboratory animals. *Laboratory Animals* **21**, 35–42.

Beynen, A.C., Baumans, V., Bertens, A.P.M.G., *et al.* (1988). Assessment of discomfort in rats with hepatomegaly. *Laboratory Animals* **22**, 320–325.

Blake, D.W., Jover, B., and McGrath, B.P. (1988) Haemodynamic and heart rate reflex responses to propofol in the rabbit. Comparison with althesin. *British Journal of Anaesthesia* **61**, 194–199.

Blouin, A., and Cormier, Y. (1987). Endotracheal intubation in guinea pigs by direct laryngoscopy. *Laboratory Animal Science* **37**, 244–245.

Bradfield, J.F., Schachtman, T.R., McLaughlin, R.M. and Steffen, E.K. (1992). Behavioural and physiological effects of inapparent wound infection in rats. *Laboratory Animal Science* **42**, 572–8.

Brammer, A., West, C.D., and Allen, S.L. (1993). A comparison of propofol with other injectable anaesthetics in a rat model for measuring cardiovascular parameters. *Laboratory Animals* **27**, 250–257.

Brearley, J.C., Kellagher, R.E.B., and Hall, L.W. (1988). Propofol anaesthesia in cats. *Journal of Small Animal Practice* **29**, 315–322.

Breckenridge, J., and Aitkenhead, A.R. (1983). Awareness during anaesthesia: a review. *Annals of the Royal College of Surgeons* **65**, 93–96.

Breivik, H. (1994). Pain management. *Clinical Anaesthesiology* **8**, 775–795.

Brody, S. (1945). "Bioenergetics and Growth". New York, Reinhold.

Brouwer, G.J., and Snowden, S.L. (1987). Breathing systems in current anaesthetic practice. *Journal of the Association of Veterinary Anaesthetists* **14**, 152–168.

Brown, B.R., and Sagalyn, A.M. (1974). Hepatic microsomal enzyme induction by inhalation anaesthetics: mechanism in the rat. *Anaesthesiology* **40**, 152–161.

Brown, L.A. (1987). Recirculation anaesthesia for laboratory fish. *Laboratory Animals* **21**, 210–215.

Brown, L.A. (1993). Anaesthesia and restraint. In: "Fish Medicine" (M.K. Stoskopf, ed.) pp. 79–90. Saunders, Philadelphia.

Buelke-Sam, J., Holson, J.F., Bazare, J.J., and Young, J.F. (1978). Comparative stability of physiological parameters during sustained anaesthesia in rats. *Laboratory Animal Science* **28** 157–162.

Bullingham, R.E.S. (1983). "Opiate Analgesia". Saunders, London.

Carruba, M.O., Bondiolotti, G.P., Picotti, G.B., Catteruccia, N., and DaPrada, M. (1987). Effects of diethyl ether, halothane, ketamine and urethane on sympathetic activity in the rat. *European Journal of Pharmacology* **134**, 15–24.

Celly, C., McDonell, W., Black, W. and Young, S. (1994). Cardiopulmonary effects of alpha2 adrenoreceptor agonists in sheep. *In* "Proceedings of the *5th International Conference of Veterinary Anaesthesia*", 117.

Child, K.J., Currie, J.P., Davis, B., Dodds, M.G., Pearce, D.R., and Twissel, D.J. (1971). The pharmacological properties in animals of CT 1341 — a new steroid anaesthetic agent. *British Journal of Anaesthesia* **43**, 2–13.

Child, K.J., Davis, B., Dodds, M.G., and Twissell, D.J. (1972a) Anaesthetic, cardiovascular and respiratory effects of a new steroidal agent CT 1341: a comparison with other intravenous anaesthetic drugs in the unrestrained cat. *British Journal of Pharmacology* **46**, 189–200.

Child, K.J., English, A.F., Gilbert, H.G., and Woollett, E.A. (1972b). An endocrinological evaluation of Althesin (CA 1341) with special reference to reproduction. *Postgraduate Medical Journal,* June supplement, 51–55.

Child, K.J., Gibson, W., Harnby, G., and Hart, J.W. (1972c) Metabolism and excretion of Althesin (CT 1341) in the rat. *Postgraduate Medical Journal,* June supplement, 37–42.

Clifford, D.H., and Soma, L.R. (1969). Feline anaesthesia. *Federation Proceedings* **28**, 1479–1499.

Cockshott, I.D., Douglas, E.J., Plummer, G.F., and Simons, P.J. (1992). The pharmacokinetics of propofol in laboratory animals. *Xenobiotica* **22**, 369–375.

Colby, E.D., and Sanford, T.D. (1981). Blood pressure and heart and respiratory rates of cats under ketamine/xylazine, ketamine/acepromazine anaesthesia. *Feline Practice* **11**, 19–24.

Cookson, J.H., and Mills, F.J. (1983). Continuous infusion anaesthesia in baboons with alphaxolone-alphadolone. *Laboratory Animals* **17**, 196–197.

Cooper, J.E. (1974) Ketamine hydrochloride as an anaesthetic for East African reptiles. *Veterinary Record* **95**, 37–41.

Cooper, J.E., and Organ, P. (1977). Pentazocine as an analgesic in dogs. *Veterinary Record* **101**, 409.

Cooper, J.E. (1984) Anaesthesia of exotic species. *In* "Manual of Anaesthesia for Small Animal Practice". (A.D.R. Hilbery, ed.) British Small Animal Veterinary Association, Cheltenham.

Costa, D.L., Lehmann, J.R., Harold, W.M. and Drew, R.T. (1986). Transcoral tracheal intubation of rodents using a fibreoptic laryngoscope. *Laboratory Animal Science* **36**, 256–261.

Coulson, N.M., Januszkiewicz, A.J., and Ripple, G.R. (1991). Physiological responses of sheep to two hours anaesthesia with diazepam–ketamine. *Veterinary Record* **129**, 329–332.

Cowan, A., Doxey, J.C., and Harry, E.J.R. (1977a). The animal pharmacology of buprenorphine, an oripavine analgesic agent. *British Journal of Pharmacology* **60**, 547–554.

Cowan, A., Lewis, J.W., and McFarlane, I.R. (1977b). Agonist and antagonist properties of buprenorphine, a new antinociceptive agent. *British Journal of Pharmacology* **60**, 537–545.

Cruz, J.I., Diaz-Otero, A., Falceto, M.W. and Cruz, A.M. (1994). Monitoring by capnography in veterinary anaesthesia in dogs and horses. *Veterinary International* **3**, 18–26.

Cunningham, F.M., and Lees, P. (1994). Advances in anti-inflammatory therapy. *British Veterinary Journal* **150**, 115–134.

Curl, J.L., and Peters, L.L. (1983). Ketamine hydrochloride and xylazine hydrochloride anaesthesia in the golden hamster (*Mesocricetus auratus*). *Laboratory Animals* **17**, 290–293.

D'Alleine, C.P., and Mann, D.D. (1982). Evaluation of ketamine/xylazine anaesthesia in the guinea pig: toxicological parameters. *Veterinary and Human Toxicology* **24**, 410–413.

DeBuf, Y. (1994). "The Veterinary Formulary — Handbook of Medicines Used in Veterinary Practice". Royal Pharmaceutical Society, 2nd edition, London.

Decker, M.J., Conrad, K.P., and Strohl, K.P. (1989). Noninvasive oximetry in the rat. *Biomedical Instrumentation and Technology*, May–June, 222–228.

Deleonardis, J.R., Clevenger, R. and Hoyt, R.F. Jr. (1995). Approaches in rodent intubation and endotracheal tube design. Contemporary Topics in Laboratory Animal Science, **34**, 60.

Desjardins, C. (1981). Endocrine signalling and male reproduction. *Biology of Reproduction* **24**, 1–21.

Dobromylskyj, P. (1993). Assessment of methadone as an anaesthetic premedicant in cats. *Journal of Small Animal Practice* **34**, 604–608.

Dodman, N.H., Clark, G.H., Court, M.H., Fikes, L.L., and Boudrieau, R.J. (1992). Epidural opioid administration for postoperative pain relief in the dog. In "Animal Pain" (C.E. Short and A. Van Poznak, eds.). pp. 274–277. Churchill Livingstone, New York.

Drummond, J.C. (1985). MAC for halothane, enflurane, and isoflurane in the New Zealand white rabbit: and a test for the validity of MAC determinations. *Anesthesiology* **62**, 336–338.

Duke, T., Komulainen Cox, A.M., Remedios, A.M., and Cribbs, P.H. (1993). The analgesic effects of administering fentanyl or medetomidine in the lumbosacral epidural space of chronically catheterised cats. *Journal of the Association of Veterinary Anaesthetists* **20**, 46.

Dum, J.E., and Herz, A.L. (1981). *In vivo* receptor binding of the opiate partial agonist, buprenorphine, correlated with its agonistic and antagonistic actions. *British Journal of Pharmacology* **74**, 627–633.

Dyson, D.H., Allen, D.G., Ingwersen, W., Pascoe, P.J., and O'Grady, M. (1987). Effects of Saffan on cardiopulmonary function in healthy cats. *Canadian Journal of Veterinary Research* **51**, 236–239.

Eales, F.A., and Small, J. (1982). Alphaxalone/alphadolone anaesthesia in the lamb. *Veterinary Record* **110**, 273–5.

Eger, E.I. (1981). Isoflurane: a review. *Anesthesiology* **55**, 559–576.

Eger, E.I. (1992). Desflurane animal and human pharmacology: aspects of kinetics, safety, and MAC. *Anesthesia and Analgesia* **75**, S3–S9.

Erhardt, W., Hebestedt, A., Aschenbrenner, G., Pichotka, B., and Blumel, G. (1984). A comparative study with various anaesthetics in mice (pentobarbitone, ketamine-xylazine, carfentanyl-etomidate). *Research in Experimental Medicine* **184**, 159–169.

Erhardt, W., Lendl, C., Hipp, R., VonHegel, G., Wiesner, G., and Wiesner, H. (1990). The use of pulse oximetry in clinical veterinary anaesthesia. *Journal of the Association of Veterinary Anaesthetists* **17**, 30–31.

Fagin, K.D., Shinsako J., and Dallman, M.F. (1983). Effects of housing and chronic cannulation on plasma ACTH and corticosterone in the rat. *American Journal of Physiology* **245**, E515–E520.

FELASA (1994). Pain and distress in laboratory rodents and lagomorphs. Report of the Federation of European Laboratory Animal Science Associations (FELASA) Working Group on Pain and Distress. *Laboratory Animals* **28**, 97–112.

Feldberg, W., and Symonds, H.W. (1980). Hyperglycaemic effect of xylazine. *Journal of Veterinary Pharmacology and Therapeutics* **3**, 197–202.

Fellows, I.W., Bastow, M.D., Byrne, A.J. and Allison, S.P. (1983). Adrenocortical suppression in multiply injured patients: a complication of etomidate treatment. *British Medical Journal* **287**, 1836–1838.

Ferstandig, L.L. (1978). Trace concentrations of anaesthetic gases: a critical review of their disease potential. *Anesthesia and Analgesia* **57**, 328–345.

Field, K.J., and Lang, C.M. (1988). Hazards of urethane (ethyl carbamate) a review of the literature. *Laboratory Animals* **22**, 255–262.

Field, K.J., White, W.J., and Lang, C.M. (1993). Anaesthetic effects of chloral hydrate, pentobarbitone and urethane in adult male rats. *Laboratory Animals* **27**, 258–269.

Field, W.E., Yelnosky, J., Mundy, J., and Mitchell, J. (1966). Use of droperidol and fentanyl for analgesia and sedation in primates. *Journal of the American Veterinary Association* **149**, 896–901.

Flecknell, P.A. (1983). Injectable anaesthetic techniques in 2 species of gerbil. *Laboratory Animals* **17**, 118–122.

Flecknell, P.A. (1984). The relief of pain in laboratory animals. *Laboratory Animals* **18**, 147–160.

Flecknell, P.A. (1991). Prevention and relief of pain and distress. In: "Animals in Biomedical Research" (C.F.M. Hendriksen, and H.B.W.M. Koeter, eds). pp. 213–234. Elsevier, Amsterdam.

Flecknell, P.A. (1994). Advances in the assessment and alleviation of pain in laboratory and domestic animals. *Journal of Veterinary Anaesthesia* **21**, 98–105.

Flecknell, P.A., and Liles, J.H. (1990). Assessment of the analgesic action of opioid agonist-antagonists in the rabbit. *Journal of the Association of Veterinary Anaesthetists* **17**, 24–29.

Flecknell, P.A., and Liles, J.H. (1991). The effects of surgical procedures, halothane anaesthesia and nalbuphine on the locomotor activity and food and water consumption in rats. *Laboratory Animals* **25**, 50–60.

Flecknell, P.A., and Liles, J.H. (1992). Evaluation of locomotor activity and food and water consumption as a method of assessing postoperative pain in rodents. In: "Animal Pain" (C.E. Short, and A. Van Poznak, eds). pp. 482–488. Churchill Livingstone, New York.

Flecknell, P.A., and Mitchell, M. (1984). Midazolam and fentanyl-fluanisone: assessment of anaesthetic effects in laboratory rodents and rabbits. *Laboratory Animals* **18**, 143–6.

Flecknell, P.A., John, M., Mitchell, M., Shurey, C., and Simpkin, S. (1983). Neuroleptanalgesia in the rabbit. *Laboratory Animals* **17**, 104–9.

Flecknell, P.A., Liles, J.H., and Wootton, R. (1989a). Reversal of fentanyl/fluanisone neuroleptanalgesia in the rabbit using mixed agonist/antagonist opioids. *Laboratory Animals* **23**, 147–155.

Flecknell, P.A., Hooper, T.L., Fetherstone, G., Locke, T.J., and McGregor, C.G.A. (1989b). Long-term anaesthesia with alfentanil and midazolam for lung transplantation in the dog. *Laboratory Animals* **23**, 278–284.

Flecknell, P.A., Liles, J.H., and Williamson, H.A. (1990a). The use of lignocaine-prilocaine local anaesthetic cream for pain-free venepuncture in laboratory animals. *Laboratory Animals* **24**, 142–146.

Flecknell, P.A., Kirk, A.J.B., Fox, C.E., and Dark, J.H. (1990b). Long-term anaesthesia with propofol and alfentanil in the dog and its partial reversal with nalbuphine. *Journal of the Association of Veterinary Anaesthetists* **17**, 11–16.

Flecknell, P.A., Kirk, A.J.B., Liles, J.H., Hayes, P.H., and Dark, J.H. (1991). Post-operative analgesia following thoracotomy in the dog: an evaluation of the effects of bupivacaine intercostal nerve block and nalbuphine on respiratory function. *Laboratory Animals* **25**, 319–324.

Flecknell, P.A., Cruz, I.J., Liles, J.H. and Whelan, G. (1996). Induction of anaesthesia with halothane and isoflurane in the rabbit: a comparison of the use of a face-mask or an anaesthetic chamber. *Laboratory Animals* (in press).

Fleischman, R.W., McCracken, D., and Forbes, W. (1977). Adynamic ileus in the rat induced by chloral hydrate. *Laboratory Animal Science* **27**, 238–243.

Frommel, E., and Joye, E. (1964). On the analgesic power of morphine in relation to age and sex of guinea-pigs. *Medicina Experimentalis* **11**, 43–46.

Gaertner, D., Boschert, K.R., and Schoeb, T.R. (1987). Muscle necrosis in Syrian hamsters resulting from intramuscular injections of ketamine and xylazine. *Laboratory Animal Science* **37**, 80–83.

Glen, J.B. (1980). Animal studies of the anaesthetic activity of ICI 35 868. *British Journal of Anaesthesia* **52**, 731–741.

Glen, J.B. (1988). Comparative aspects of equipment for the intravenous infusion of anaesthetic agents. *Journal of the Association of Veterinary Anaesthetists* **15**, 65–79.

Glen, J.B., and Hunter, S.C. (1984). Pharmacology of an emulsion formulation of ICI 35 868. *British Journal of Anaesthesia* **56**, 617–626.

Glynn, C.J. (1987). Intrathecal and epidural administration of opiates. *Clinical Anaesthesiology* **1**, 915–933.

Green, C.J. (1975). Neuroleptanalgesic drugs combinations in the anaesthetic management of small laboratory animals. *Laboratory Animals* **9**, 161–78.

Green, C. (1981a). Anaesthetic gases and health risks to laboratory personnel: a review. *Laboratory Animals* **15**, 397–403.

Green, C.J. (1981b). "Animal Anaesthesia''. Theyden Bois, Laboratory Animals Ltd, Essex.

Green, C.J., Halsey, M.J., Precious, S., and Simpkin, S. (1978). Alphaxolone-alphadolone anaesthesia in laboratory animals. *Laboratory Animals* **12**, 85–89.

Green, C.J., Knight, J., Precious, S., and Simpkin, S. (1981a). Ketamine alone and combined with diazepam or xylazine in laboratory animals: a 10-year experience. *Laboratory Animals* **15**, 163–170.

Green, C.J., Knight, J., Precious, S., and Wardley-Smith, B. (1981b). Metomidate, etomidate and fentanyl as injectable anaesthetic agents in mice. *Laboratory Animals* **15**, 171–175.

Greene, S.A., and Thurmon, J.C. (1988). Xylazine — a review of its pharmacology and use in veterinary medicine. *Journal of Veterinary Pharmacology and Therapeutics* **11**, 295–313.

Hall, G.M. (1985). The anaesthetic modification of the endocrine and metabolic response to surgery. *Annals of the Royal College of Surgeons of England* **67**, 25–29.

Hall, L.W., and Chambers, J.P. (1987). A clinical trial of propofol infusion anaesthesia in dogs. *Journal of Small Animal Practice* **28**, 623–637.

Hall, L.W. and Clarke, K.W. (1991). "Veterinary Anaesthesia", 9th Ed. Baillière Tindall, London.

Hall, L.W., and Taylor, P.M. (1994). "Anaesthesia of the Cat". Baillière Tindall, London.

Harrison, G.J., and Harrison, L.R. (1986). "Clinical Avian Medicine and Surgery". Saunders, Philadelphia.

Hart, M.V., Rowles, J.R., Hohimer, A.R., Morton, M.J., and Hosenpud, J.D. (1984). Hemodynamics in the guinea pig after anaesthetization with ketamine/xylazine. *American Journal of Veterinary Research* **45**, 2328–2330.

Heel, R.C., Brogden, R.N., Speight, T.M., and Avery, G.S. (1979). Buprenorphine; a review of its pharmacological properties and therapeutic efficacy. *Drugs* **17**, 81–110.

Hensley, F.A. and Martin, D.E. (1990). "The Practice of Cardiac Anaesthesia". Little, Brown, Boston.

Hermansen, K., Pedersen, L.E., and Olesen, H.O. (1986). The analgesic effect of buprenorphine, etorphine and pethidine in the pig: a randomized double blind cross-over study. *Acta Pharmacologica et Toxicologica* **59**, 27–35.

Hird, J.F.R., and Carlucci, F. (1977). The use of the co-axial circuit to control the degree of rebreathing in the anaesthetized dog. *Proceedings of the Association of Veterinary Anaesthetists of Great Britain and Ireland* **7**, 13–21.

Holzgrefe, H.H., Everitt, J.M., and Wright, E.M. (1987). Alpha-chloralose as a canine anaesthetic. *Laboratory Animal Science* **37**, 587–593.

Hoover-Plow, J.L., and Clifford, A.J. (1978). The effect of surgical trauma on muscle protein turnover in rats. *Biochemistry Journal* **176**, 137–142.

Hsu, W.H., Bellin, S.I., Dellman, H.D., Habil, V., and Hanson, C.E. (1986). Xylazine-ketamine-induced anaesthesia in rats and its antagonism by yohimbine. *Journal of the American Veterinary Medical Association* **189**, 1040–1043.

Hu, C., Flecknell, P.A., and Liles, J.H. (1992). Fentanyl and medetomidine anaesthesia in the rat and its reversal using atipamazole and either nalbuphine or butorphanol. *Laboratory Animals* **26**, 15–22.

Hughes, E.W., Martin-Body, R.L., Sarelius, I.H., and Sinclair, J.D. (1982). Effects of urethane–chloralose anaesthesia on respiration in the rat. *Clinical and Experimental Pharmacology and Physiology* **9**, 119–127.

Hughes, P.J., Doherty, M.M., and Charman, W.N. (1993). A rabbit model for the evaluation of epidurally administered local anaesthetic agents. *Anaesthesia and Intensive Care* **21**, 298–303.

Hunter, S.C., Glen, J.B., and Butcher, C.J. (1984). A modified anaesthetic vapour extraction system. *Laboratory Animals* **18**, 42–44.

IASP (International Association For the Study of Pain) (1979). "Report of Subcommittee on taxonomy". *Pain* **6**, 249–252.

ILAR (Institute of Laboratory Animal Resources) (1992). "Recognition and Alleviation of Pain and Distress in laboratory animals". National Academy Press, Washington.

Ingwersen, W., Allen, D.G., Dyson, D.H., Pascoe, P.J., and O'Grady, M.R. (1988). Cardiopulmonary effects of a ketamine hydrochloride/acepromazine combination in healthy cats. *Canadian Journal of Veterinary Research* **52**, 1–4.

Iwama, G.K. (1992). Anaesthesia, analgesia and euthanasia in fish. *In* "The care and use of amphibians, reptiles and fish in research" (D.O. Schaeffer, K.M. Kleinow, and L. Krulisch, eds). pp. 167–174 Scientists Centre for Animal Welfare, Bethesda, MD.

Jacobson, J.D., Miller, M.W., Matthews, N.S., Hartsfield, S.M., and Knauer, K.W. (1992). Evaluation of accuracy of pulse oximetry in dogs. *American Journal of Veterinary Research* **53**, 537–540.

Janssen, P.A.J., Niemegeers, C.J.E., and Marsboom, R.P.H. (1975). Etomidate, a potent non-barbiturate hypnotic. Intravenous etomidate in mice, rats, guinea-pigs, rabbits and dogs. *Archives of International Pharmacodynamics* **214**, 92–132.

Jolly, D.W., Mawdesley-Thomas, L.E., and Bucke, D. (1972). Anaesthesia of fish. *Veterinary Record* **91**, 424–426.

Kain, M.L., and Nunn, J.F. (1968). Fresh gas economics of the Magill circuit. *Anesthesiology* **29**, 964–974.

Keeri-Szanto, M. (1983). Demand analgesia. *British Journal of Anaesthesia* **55**, 919–920.

Kehlet, H. (1978). Influence of epidural analgesia on the endocrine metabolic response to surgery. *Acta Anaesthesiologica Scandinavica* (suppl.) **70**, 39–42.

Kero, P., Thomasson, B., and Soppi, A.M. (1981). Spinal anaesthesia in the rabbit. *Laboratory Animals* **15**, 347–348.

Kissin, I., Motomura, S., Aultman, D.F., and Reves, J.G. (1983). Inotropic and anaesthetic potencies of etomidate and thiopental in dogs. *Anesthesia and Analgesia* **62**, 961–965.

Kistler, P. (1988). *Zur Schmerzbekampfung im Tierversuch (Attenuation of pain in animal experimentation).* PhD Thesis, University of Bern.

Kittleson, M.D., and Olivier, N.B. (1983). Measurement of systemic arterial blood pressure. *Veterinary Clinics of North America: Small Animal Practice* **13**, 321–336.

Klide, A.M., and Soma, L.R. (1968). Epidural analgesia in the dog and cat. *Journal of the American Veterinary Medical Association* **153**, 165–173.

Ko, J.C.H., Thurmon, J.C., Tranquilli, W.J., Benson, G.J., and Olson, W.A. (1992). A comparison of medetomidine–propofol and medetomidine–midazolam–propofol anaesthesia in rabbits. *Laboratory Animal Science* **42**, 503–507.

Koblin, D.D. (1992). Characteristics and implications of desflurane metabolism and toxicity. *Anesthesia and Analgesia* **75**, S10–S16.

Korner, P.I., Uther, J.B., and White, S.W. (1968). Circulatory effects of chloralose-urethane and sodium-pentobarbitone anaesthesia in the rabbit. *Journal of Physiology* **199**, 253–65.

Kruse-Elliott, K.T., Swanson, C.R., and Aucoin, D.P. (1987). Effects of etomidate on adrenocortical function in canine surgical patients. *American Journal of Veterinary Research* **48**, 1098–1100.

Laitinen, O.M. (1990). Clinical observations on medetomidine/ketamine anaesthesia in sheep and its reversal by atipamezole. *Journal of the Association of Veterinary Anaesthetists* **17**, 17–19.

LASA (Laboratory Animal Science Association) (1990). The assessment and control of the severity of scientific procedures on laboratory animals. *Laboratory Animals* **24**, 97–130.

Latasch, L., Probst, S., and Dudziak, R. (1984). Reversal by nalbuphine of respiratory depression caused by fentanyl. *Anesthesia and Analgesia* **63**, 814–816.

Lees, P., May, S.A., and McKellar, Q.A. (1991). Pharmacology and therapeutics of non-steroidal anti-inflammatory drugs in the dog and cat. 1: General pharmacology. *Journal of Small Animal Practice* **32** 183–193.

Leese, T., Husken, P.A., and Morton, D.B. (1988). Buprenorphine analgesia in a rat model of acute pancreatitis. *Surgical Research Communications* **3**, 53–60.

Liles, J.H. (1994). *The assessment and alleviation of pain in laboratory rats.* PhD Thesis, Newcastle University.

Liles, J.H., and Flecknell, P.A. (1992). The use of non-steroidal anti-inflammatory drugs for the relief of pain in laboratory rodents and rabbits. *Laboratory Animals* **26**, 241–255.

Liles, J.H., and Flecknell, P.A. (1993a). A comparison of the effects of buprenorphine, carprofen and flunixin following laparotomy in rats. *Journal of Veterinary Pharmacology and Therapeutics* **17**, 284–290.

Liles, J.H., and Flecknell, P.A. (1993b). The effects of surgical stimulus on the rat and the influence of analgesic treatment. *British Veterinary Journal* **149**, 515–525.

Linde, H.W., and Berman, M.L. (1971). Nonspecific stimulation of drug metabolizing enzymes by inhalation anesthetic agents. *Anesthesia and Analgesia (current researches)* **50**, 656–665.

Lipman, N.S., Phillips, P.A., and Newcomer, C.E. (1987). Reversal of ketamine/xylazine anaesthesia in the rabbit with yohimbine. *Laboratory Animal Science* **37**, 474–477.

Lipman, N.S., Marini, R.P., and Erdman, S.E. (1990). A comparison of ketamine/xylazine and ketamine/xylazine/acepromazine anaesthesia in the rabbit. *Laboratory Animal Science* **40**, 395–398.

Lovell, D.P. (1986a). Variation in barbiturate sleeping time in mice. 3. Strain × environment interactions. *Laboratory Animals* **20**, 307–312.

Lovell, D.P. (1986b). Variation in pentobarbitone sleeping time in mice. 1. Strain and sex differences. *Laboratory Animals* **20**, 85–90.

Lovell, D.P. (1986c). Variation in pentobarbitone sleeping time in mice. 2. Variables affecting test results. *Laboratory Animals* **20**, 91–96.

Lumb, W.V., and Wynn Jones, E. (1973). "Veterinary Anaesthesia". Lee & Febiger, Philadelphia.

Mahmoudi, N.W., Cole, D.J. and Shapiro, H.M. (1989). Insufficient anaesthetic potency of nitrous oxide in the rat. *Anaesthesiology* **70**, 345–349.

Manley, S.V., and McDonnell, W.N. (1979a). A new circuit for small animal anaesthesia: the Bain coaxial circuit. *Journal of the American Animal Hospital Association* **15**, 61–65.

Manley, S.V., and McDonnell, W.N. (1979b). Clinical evaluation of the Bain breathing circuit in small animal anaesthesia. *Journal of the American Animal Hospital Association* **15**, 67–72.

Marietta, M.P., White, P.F., Pudwill, C.R., Way, W.L., and Trevor, A.J. (1975). Biodisposition of ketamine in the rat: self-induction of metabolism. *Journal of Pharmacology and Experimental Therapeutics* **196**, 536–544.

Marini, R., Avison, D.L., Corning, B.F., and Lipman, N.S. (1992). Ketamine/xylazine/butorphanol: a new anaesthetic combination for rabbits. *Laboratory Animal Science* **42**, 57–62.

Marini, R.P., Hurley, R.J., Avison, D.L., and Lipman, N.S. (1993). An evaluation of three neuroleptanalgesic combinations in rabbits. *Laboratory Animal Science* **43**, 338–345.

Martin, L. (1992). "All You Really Need to Know to Interpret Arterial Blood Gases". Lea & Febiger, Philadelphia.

Mather, L.G. (1983). Pharmacokinetic and pharmacodynamic factors influencing the choice, dose and route of administration of opiates for acute pain. *Clinics in Anesthesiology* **1**, 17–40.

Mathews, K.A., Doherty, T., Dyson, D.H., Wilcock, B., and Valliant, A. (1990). Nephrotoxicity in dogs associated with methoxyflurane anaesthesia and flunixin meglumine analgesia. *Canadian Veterinary Journal* **31**, 766–771.

Mathews, K.A., Paley, D., Binnington, A., Young, S. and Foster, R. (1994). A comparison of NSAIDs vs opioids in relieving post-operative pain. *In* "Proceedings of the *International Congress of Veterinary Anaesthesia*". p. 169. University of Guelph.

Mazze, R.I., Rice, S.A., and Baden, J.M. (1985). Halothane, isoflurane, and enflurane MAC in pregnant and nonpregnant female and male mice and rats. *Anaesthesiology* **62**, 339–341.

Mburu, D.N., Mbugua, S.W., Skoglund, L.A., and Lokken, P. (1988). Effects of paracetamol and acetylsalicylic acid on the post-operative course after experimental orthopaedic surgery in dogs. *Journal of Veterinary Pharmacology and Therapeutics* **11**, 163–171.

McGlone, J.J., and Hellman, J.M. (1988). Local and general anaesthetic effects on behaviour and performance of two- and seven-week old castrated and uncastrated piglets. *Journal of Animal Science* **66**, 3049–3058.

McNeil, P.E. (1992). Acute tubulo-interstitial nephritis in a dog after halothane anaesthesia and administration of flunixin, meglumine and trimethoprim-sulphate. *Veterinary Record* **131**, 148–151.

McQuay, H.J., Carroll, D., and Moore, R.A. (1988). Post-operative orthopaedic pain — the effect of opiate premedication and local anaesthetic blocks. *Pain* **33**, 291–295.

Michalot, G., Girardet, P., Grimbert, F., Piasentin, D., and Stieglitz, P. (1980). 24-hour Althesin-fentanyl anaesthesia in dogs. Time course of haemodynamic changes. *British Journal of Anaesthesia* **52**, 19–22.

Middleton, D.J., Ilkiw, J.E., and Watson, A.D.J. (1982). Physiological effects of thiopentone, ketamine and CT 1341 in cats. *Research in Veterinary Science* **32**, 157–162.

Morgan, D.W.T., and Legge, K. (1989). Clinical evaluation of propofol as an intravenous anaesthetic agent in cats and dogs. *Veterinary Record* **124**, 31–33.

Morris, T.H. (1991). Use of medetomidine and atipamezole in laboratory animals. *In*: "Proceedings of the *4th International Congress of Veterinary Anaesthesia*". *Journal of Veterinary Anaesthesia*, 277–279.

Morris, T.H. (1995). Antibiotic therapeutics in laboratory animals. *Laboratory Animals* **29**, 16–36.

Morris, T.H., Jonsson, J., Kokkenen, U-M., *et al.* (1994). Selection of anaesthetics for experimental procedures using animals. *In* "Proceedings of the *SCANDLAS Symposium*". Bergen, 1994.

Morton, D.B., and Griffiths, P.H.M. (1985). Guidelines on the recognition of pain, distress and discomfort in experimental animals and an hypothesis for assessment. *Veterinary Record* **116**, 431–436.

Mulder, K.J., and Mulder, J.B. (1979). Ketamine and xylazine anaesthesia in the mouse. *Veterinary Medicine and Small Animal Clinician* **74**, 569–570.

Murray, W.J., and Fleming, P.J. (1972). Defluorination of methoxyflurane during anaesthesia: comparison of man with other species. *Anaesthesiology* **37**, 620–625.

Nagel, M.L., Muir, W.W., and Nguyen, I.C. (1979). Comparison of the cardiopulmonary effects of etomidate and thiamylal in dogs. *American Journal of Veterinary Research* **40**, 193–196.

Nevalainen, T., Pyhala, L., Voipio, H.-M., and Virtanen, R. (1989). Evaluation of anaesthetic potency of medetomidine-ketamine combination in rats, guinea-pigs and rabbits. *Acta Veterinaria Scandinavica* **85**, 139–143.

Nolan, A., and Reid, J. (1993). Comparison of the post-operative analgesic and sedative effects of carprofen and papaveretum in the dog. *Veterinary Record* **133**, 240–242.

Nolan, A., Livingstone, A., and Waterman, A. (1987). Investigation of the antinociceptive activity of buprenorphine in sheep. *British Journal of Pharmacology* **92**, 527–533.

Norris, M. (1981). Portable anaesthetic apparatus designed to induce and maintain surgical anaesthesia by methoxyflurane inhalation in the Mongolian gerbil (*Meriones unguiculatus*). *Laboratory Animals* **15**, 153–155.

Norris, M.L., and Turner, W.D. (1983). An evaluation of tribromoethanol (TBE) as an anaesthetic agent in the Mongolian gerbil (*Meriones unguiculatus*). *Laboratory Animals* **15**, 153–155.

Nowich, K.H. (1977). Molecular dynamics in biosystems. The kinetics of tracers in intact organisms. Pergamon, New York.

O'Hair, K.C., Dodd, K.T., Phillips, Y.Y., and Beattie, R.J. (1988). Cardiopulmonary effects of nalbuphine hydrochloride and butorphanol tartrate in sheep. *Laboratory Animal Science* **38**, 58–61.

Olson, M.E., Vizzutti, D., Morck, D.W., and Cox, A.K. (1993). The parasympatholytic effects of atropine sulphate and glycopyrrolate in rats and rabbits. *Canadian Journal of Veterinary Research* **57**, 254–258.

Pablo, L.S. (1993). Epidural morphine in goats after hindlimb orthopaedic surgery. *Veterinary Surgery* **22**, 307–310.

Palminteri, A. (1963). Oxymorphone an effective analgesic in dogs and cats. *Journal of American Veterinary Medical Association* **143**, 160–163.

Papaioannou, V.E., and Fox, J.G. (1993). Efficacy of tribromoethanol anaesthesia in mice. *Laboratory Animal Science* **43**, 189–192.

Park, C.M., Clegg, K.E., Harvey-Clark, C.J. and Hollenberg, M.J. (1992). Improved techniques for successful neonatal rat surgery. *Laboratory Animal Science* **42**, 508–513.

Pascoe, P.J. (1993). Analgesia after lateral thoracotomy in dogs. Epidural morphine vs. intercostal bupivacaine. *Veterinary Surgery* **22**, 141–147.

Peeters, M., Gil, D., Teske, E., *et al.* (1988). Four methods for general anaesthesia in the rabbit: a comparative study. *Laboratory Animals* **22**, 355–360.

Pekow, C. (1992). Buprenorphine Jell-O recipe for rodent analgesia. *Synapse* **25**, 35–36.

Phillips, I.R., and Grist, S.M. (1975). Clinical use of CT.1341 anaesthetic ('Saffan') in marmosets. *Laboratory Animals* **9**, 57–60.

Pick, C., Cheng, J., Paul, D., and Pasternak, G.W. (1991). Genetic influences in opioid analgesic sensitivity in mice. *Brain Research* **566**, 295–298.

Pieri, L., Schaffner, R., and Scherschlicht, R. *et al.* (1981). Pharmacology of Midazolam. *Arzneim-Forsch/Drug Research* **31**, 2180–2201.

Pircio, A.W., Gylys, J.A., Cavanagh, R.L., Buyniski, J.P., and Bierwagen, M.E. (1976). The pharmacology of butorphanol, a 3, 14-dihydroxymorphinan narcotic antagonist analgesic. *Archives Internationales de Pharmacodynamie et de Therapie* **220**, 231–257.

Popilskis, S., Kohn, D.F., Laurent, L., and Danilo, P. (1993). Efficacy of epidural morphine versus intravenous morphine for post-thoracotomy pain in dogs. *Journal of Veterinary Anaesthesia* **20** (June), 21–25.

Popio, K.A., Jackson, D.H., Ross, A.M., Schreiner, B.F., and Yu, P.N. (1978). Hemodynamic and respiratory effects of morphine and butorphanol. *Clinical Pharmacology and Therapeutics* **23**, 281–287.

Rees, G.J. (1950). Anaesthesia in the newborn. *British Medical Journal* **2**, 1419–1422.

Regan, M.J., and Eger, E.I. (1967). The effect of hypothermia in dogs on anaesthetizing and apnoeic doses of inhalation agents. *Anesthesiology* **28**, 689–700.

Reid, J., and Nolan, A.M. (1991). A comparison of the postoperative analgesic and sedative effects of flunixin and papaveretum in the dog. *Journal of Small Animal Practice* **32**, 603–608.

Remie, R., Bertens, A.P.M.G., Van Dongen, J.W., Rensema, J.W., and Van Wunnik, G.H.J. (1990).

Anaesthesia of the laboratory rat. *In* "Manual of Microsurgery on the Laboratory Rat" (J.W. Van Dongen, J.W. Rensema, and G.H.J. Van Wunnik, eds). pp. 61–80 Elsevier, Amsterdam.

Robertson, D.H., and Laing, A.W. (1980). Intravenous buprenorphine (temgesic): use following fentanyl analgesic anaesthesia. *Clinical Trials Journal* **17**, 51–55.

Robertson, S.A., Johnston, S., and Beemsterboer, J. (1992). Cardiopulmonary, anaesthetic, and post-anaesthetic effects of intravenous infusions of propofol in greyhounds and non-greyhounds. *American Journal of Veterinary Research* **53**, 1027–1032.

Robinson, F.P., and Patterson, C.C. (1985). Changes in liver function tests after propofol ('Diprivan'). *Postgraduate Medical Journal* (suppl. 3) **61**, 160–161.

Rosow, C.E. (1985). New synthetic opioid analgesics. *In* "Acute Pain" (G. Smith, and B.G. Covino, eds). pp. 68–103. Butterworths, London.

Rubal, B., and Buchanan, C. (1986). Supplemental chloralose anaesthesia in morphine premedicated dogs. *Laboratory Animal Science* **36**, 59–64.

Russell, W.M.S., and Burch, R.L. (1992). "The Principles of Humane Experimental Technique". UFAW, Potters Bar, Herts.

Sainsbury, A.W., Eaton, B.D., and Cooper, J.E. (1991). An investigation into the use of propofol (Rapinovet) in long-tailed macaques (*Macaca fascicularis*). *Journal of Veterinary Anaesthesia* **18**, 38–41.

Salo, M. (1988). The relevance of metabolic and endocrine responses to anaesthesia and surgery. *Acta Anaesthesiologica Belgica* **39**, 133–141.

Samour, J.H., Jones, D.M., Knight, J.A. and Howlett, J.C. (1984). Comparative studies of the use of some injectable anaesthetic agents in birds. *Veterinary Record* **115**, 6–11.

Sawyer, D., and Rech, R.H. (1987). Analgesia and behavioural effects of butorphanol nalbuphine, and pentazocine in the cat. *Journal of the American Animal Hospital Association* **23**, 438–446.

Sawyer, D., Rech, R.H., Durham, R.A., Adams, T., Richter, M.A., and Striler, E.L. (1991). Dose response to butorphanol administered subcutaneously to increase visceral nociceptive threshold in dogs. *American Journal of Veterinary Research* **52**, 1826–1830.

Sawyer, D.G., Rech, R.H., Adams, T., Durham, R.A., Richter, M.A., and Striler, E.L. (1992). Analgesia and behavioural responses of dogs given oxymorphone-acepromazine and meperidine-acepromazine after methoxyflurane and halothane anaesthesia. *American Journal of Veterinary Research* **53**, 1361–1368.

Schaeffer, D.O., Kleinow, K.M., and Krulisch, L. (1992). "The Care and Use of Amphibians, Reptiles and Fish in Research". Scientists Centre for Animal Welfare, Bethesda, MD.

Sear, J.W., Uppington, J., and Kay, N.H. (1985). Haematological and biochemical changes during anaesthesia with propofol ('Diprivan'). *Postgraduate Medical Journal* (suppl. 3) **61**, 165–168.

Sebel, P.S., and Lowdon, J.D. (1989). Propofol: a new intravenous anaesthetic. *Anesthesiology* **71**, 260–277.

Sebesteny, A. (1971). Fire-risk-free anaesthesia of rodents with halothane. *Laboratory Animals* **5**, 225–231.

Sharp, F.R., and Hammel, H.T. (1972). Effects of chloralose-urethan anaesthesia on temperature-regulation in dogs. *Journal of Applied Physiology* **33**, 229–233.

Short, C.E. (1987). "Principles and Practice of Veterinary Anaesthesia". Williams and Wilkins, Baltimore.

Short, C.E., and Van Poznak, A. (1992). "Animal Pain." Churchill Livingstone, New York.

Shukla, R., and Shukla, S.B. (1983). The effect of polyethylene glycol-200 on metabolic acidosis induced by chloralose anaesthesia. *Journal of Veterinary Pharmacology and Therapeutics* **6**, 149–152.

Sisson, D., and Siegel, J. (1989). Chloral hydrate anaesthesia: EEG power spectrum analysis and effects on VEPs in the rat. *Neurotoxicology and Teratology* **11**, 51–56.

Skolleborg, K.C., Gronbech, J.E., Grong, K., Abyholm, F.E., and Lekven, J. (1990). Distribution of cardiac output during pentobarbital versus midazolam/fentanyl/fluanisone anaesthesia in the rat. *Laboratory Animals* **24**, 221–227.

Skues, M.A., Watson, D.M., O'Meara, M., and Goddard, J.M. (1993). Patient-controlled analgesia in children. A comparison of two infusion techniques. *Paediatric Anaesthesia* **3**, 223–228.

Smiler, K.L., Stein, S., Hrapkiewicz, K.L., and Hiben, J.R. (1990). Tissue response to intramuscular and intraperitoneal injections of ketamine and xylazine in rats. *Laboratory Animal Science* **40**, 60–64.

Smith, G. (1984). Postoperative pain. *In* "Quality of Care in Anaesthetic Practice", (J.N. Lunn, ed.) pp. 164–192. Macmillan, London.

Smith, G., and Covino, B.G. (1985). "Acute Pain". Butterworths, London.

Stark, R.D., Binks, S.M., Dutka, V.N., O'Connor, K.M., Arnstein, M.J.A., and Glen, J.B. (1985). A review of the safety and tolerance of propofol ('Diprivan'). *Postgraduate Medical Journal* (suppl. 3) **61**, 152–156.

Steffey, E., Gillespie, J.R., Berry, J.D., Eger-II, E.I., and Munson, E.S. (1974). Anaesthetic potency (MAC) of nitrous oxide in the dog, cat and stump-tail monkey. *Journal of Applied Physiology* **36**, 530–532.

Steffey, E.P. (1994). Inhalation anaesthesia. *In:* "Anaesthesia of the Cat" (L.W. Hall and P.M. Taylor, eds). pp. 157–193. Bàlliére Tindall, London.

Svendsen, P., Ainsworth, M., and Carter, A. (1990). Acid-base status and cardiovascular function in pigs anaesthetized with α-chloralose. *Scandinavian Journal of Laboratory Animal Science* **17**, 89–95.

Taylor, P.M. (1985). Analgesia in the dog and cat. *In Practice* **17**, 5–13.

Taylor, P.M. (1991). Anaesthesia in sheep and goats. *In Practice* **13**, 31–36.

Taylor, P.M. (1992). Use of xylocaine pump spray for intubation in cats. *Veterinary Record* **130**, 583.

Taylor, P., and Herrtage, M.E. (1986). Evaluation of some drug combinations for sedation in the dog. *Journal of Small Animal Practice* **27**, 325–333.

Taylor, P.M., and Houlton, J.E.F. (1984). Postoperative analgesia in the dog: comparison of morphine, buprenorphine and pentazocine. *Journal of Small Animal Practice* **25**, 437–451.

Thomasson, B., Ruuskanen, O., and Merikanto, J. (1974). Spinal anaesthesia in the guinea pig. *Laboratory Animals* **8**, 241–244.

Tranquilli, W.J., Thurmon, J.C., and Benson, G.J. (1985). Anaesthetic potency of nitrous oxide in young swine (*Sus scrofa*). *American Journal of Veterinary Research* **46**, 58–60.

Tranquilli, W.J., Thurmon, J.C., Speiser, J.R., Benson, G.J., and Olson, W.A. (1988). Butorphanol as a preanaesthetic in cats: its effects on two common intramuscular regimens. *Veterinary Medicine* **83**, 848–854.

Trim, C.M. (1983). Cardiopulmonary effects of butorphanol tartrate in dogs. *American Journal of Veterinary Research* **44**, 329–331.

Trim, C.M. (1987). Sedation and Anaesthesia. *In* "Diseases of the Cat', Medicine and Surgery" Vol. 1. (J. Holzworth, ed.) W.B. Saunders, Philadelphia.

Tulamo, R.-M., Raekallio, M., and Ekblad, A. (1995). Cardiovascular effects of medetomidine-ketamine anaesthesia in sheep, with and without 100% oxygen and its reversal with atipamezole. *Proceedings of the Association of Veterinary Anaesthetists.* 71.

Ungercr, M. (1978). A comparison between the Bain and Magill anaesthetic systems during spontaneous breathing. *Canadian Anesthesia Society Journal* **25**, 122–125.

Van-Pelt, L.F. (1977). Ketamine and xylazine for surgical anaesthesia in rats. *Journal of the American Veterinary Medical Association* **171**, 842–844.

Vanacker, B., Dekegel, D., Dionys, J., *et al.* (1987). Changes in intraocular pressure associated with the administration of propofol. *British Journal of Anaesthesia* **59**, 1514–1517.

Vegfors, M., Sjoberg, F., Lindberg, L.-G., Gustafsson, U., and Lennmarken, C. (1991). Basic studies of pulse oximetry in a rabbit model. *Acta Anaesthesiologica Scandinavica* **35**, 596–599.

Verstegen, J., Fargetton, X., and Ectors, F. (1989). Medetomidine/ketamine anaesthesia in cats. *Acta Veterinaria Scandinavica* **85**, 117–123.

Verstegen, J., Fargetton, X., Donnay, I., and Ectors, F. (1990). Comparison of the clinical utility of medetomidine/ketamine and xylazine/ketamine combinations for the ovariectomy of cats. *Veterinary Record* **127**, 424–426.

Verstegen, J., Fargetton, X., Donnay, I., and Ectors, F. (1991a). An evaluation of medetomidine/ketamine and other drug combinations for anaesthesia in cats. *Veterinary Record* **128**, 32–35.

Verstegen, J., Fargetton, X., Zanker, S., Donnay, I., and Ectors, F. (1991b). Antagonistic activities of atipamezole, 4-aminopyridine and yohimbine against medetomidine/ketamine-induced anaesthesia in cats. *Veterinary Record* **128**, 57–60.

Virtanen, R. (1989). Pharmacological profiles of medetomidine and its antagonist, atipamezole, *Acta Veterinaria Scandinavica* **4**, 29–37.

Voipio, H.M., Nevalainen, T.O., and Virtanen, R. (1990). Evaluation of anaesthetic potency of medetomidine-ketamine combination in mice. *IX*th *ICLAS Symposium Proceedings*, 298–299.

Wagner, J.G. (1974). A safe method for rapidly achieving plasma concentration plateaus. *Clinical Pharmacology and Therapeutics* **16**, 691–700.

Wagner, R., and White, P.F. (1984). Etomidate inhibits adrenocortical function in surgical patients. *Anesthesiology* **61**, 647–651.

Walker, L.A., Buscemi-Bergin, M., and Gellai, M. (1983). Renal-hemodynamics in conscious rats: effects of anesthesia, surgery and recovery. *American Journal of Physiology* **245**, F67–F74.

Waterman, A. (1988). Use of propofol in sheep. *Veterinary Record* **122**, 260.

Watkins, S.B., Hall, L.W., and Clarke, K.W. (1987). Propofol as an intravenous anaesthetic agent in dogs. *Veterinary Record* **120**, 326–329.

Weaver, B.M.Q., and Raptopoulos, D. (1990). Induction of anaesthesia in dogs and cats with propofol. *Veterinary Record* **126**, 617–620.

Weiskopf, R.B., and Bogetz, M.S. (1984). Minimum alveolar concentrations (MAC) of halothane and nitrous oxide in swine. *Anesthesia and Analgesia* **63**, 529–32.

Whelan, G., and Flecknell, P.A. (1992). The assessment of depth of anaesthesia in animals and man. *Laboratory Animals* **26**, 153–162.

Whelan, G., and Flecknell, P.A. (1994). The use of etorphine/methotrimeprazine and midazolam as an anaesthetic technique in laboratory rats and mice. *Laboratory Animals* **28**, 70–77.

Whelan, G., and Flecknell, P.A. (1995). Anaesthesia of laboratory rabbits using etorphine/methotrimeprazine and midazolam. *Laboratory Animals* **29**, 83–89.

White, G.L., and Cunnings, J.F. (1979). A comparison of ketamine and ketamine/xylazine in the baboon *Veterinary Medicine and Small Animal Clinician* **74**, 392–396.

White, G.L., and Holmes, D.D. (1976). A comparison of ketamine and the combination of ketamine-xylazine for effective surgical anesthesia in the rabbit. *Laboratory Animal Science* **26**, 804–806.

White, P.F., Way, W.L., and Trevor, A.J. (1982). Ketamine — its pharmacology and therapeutic uses *Anesthesiology* **56**, 119–136.

Wiklund, L., and Thoren, L. (1985). Intraoperative blood component and fluid therapy. *Acta Anaesthesiologica Scandinavica 29* **1–8.**

Wixson, S.K., White, W.J., Hughes, H.C., Marshall, W.K., and Lang, C.M. (1987). The effects of pentabarbital, fentanyl–droperidol, ketamine–xylazine and ketamine–diazepam on noxious stimulus perception in adult male rats. *Laboratory Animal Science* **37**, 731–735.

Wolfensohn, S., and Lloyd, M. (1994). "Handbook of Laboratory Animal Management and Welfare" Oxford University Press.

Wood, M., and Wood, A.J.J. (1984). Contrasting effects of halothane, isoflurane and enflurane on *in-vivo* drug metabolism in the rat. *Anesthesia and Analgesia* **63**, 709–714.

Wood, G.N., Molony, V., and Fleetwood-Walker, S.M. (1991). Effects of local anaesthesia and intravenous naloxone on the changes in behaviour and plasma concentrations of cortisol produced by castration and tail docking with tight rubber rings in young lambs. *Research in Veterinary Science* **51**, 193–199.

Woolf, C.J., and Wall, P.D. (1986). Morphine sensitive and morphine insensitive actions of C-fibre input on the rat spinal cord. *Neuroscience Letters* **64**, 211–225.

Wright, M. (1982). Pharmacologic effects of ketamine and its use in veterinary medicine. *Journal of the American Veterinary Medical Association* **180**, 1462–1471.

Yaksh, T.L., Al-Rodhan, N.R.F., and Mjanger, E. (1988). Sites of action of opiates in production of analgesia. *In* "Anaesthesia Review 5" (L. Kaufman, ed.). pp. 254–268. Churchill Livingstone, London.

Yoxall, A.T. (1978). Pain in small animals — its recognition and control. *Journal of Small Animal Practice* **19**, 423–438.

APPENDIX 1

Recommended Techniques and Physiological Data
(When no injectable anaesthetic is recommended, inhalation agents should be used)

	Rat	Mouse
Adult body weight	250–350 g	30–40 g
Body temperature	38°C	37·4°C
Respiration rate	80 min^{-1}	180 min^{-1}
Resting heart rate	350 min^{-1}	570 min^{-1}
Sedation/pre-medication	Fentanyl/fluanisone (Hypnorm) (0·5 ml kg^{-1} i/p)	Fentanyl/fluanisone (Hypnorm) (0·4 ml kg^{-1} i/p)
Injectable anaesthesia		
Short-term (5–15 min)	Propofol (10–12 mg kg^{-1} i/v)	Propofol (26 mg kg^{-1} i/v)
Medium term (20–30 min)	Fentanyl/fluanisone and midazolam (2·7 ml kg^{-1} i/p of 2 parts water for injection, 1 part Hypnorm and 1 part midazolam)	Fentanyl/fluanisone and midazolam (10 ml kg^{-1} i/p of 2 parts water for injection, 1 part Hypnorm and 1 part midazolam)
Long term (1–12 h)	Alphaxalone/alphadolone (10–12 mg kg^{-1} i/v) then 0·2–0·7 mg kg^{-1} min^{-1}). Alternatively, continue medium-term regimen with fentanyl/fluanisone (0·1–0·2 ml kg^{-1} i/m every 30–40 min) and midazolam (2 mg kg^{-1} i/p, every 4 hours)	Alphaxalone/alphadolone (15–20 mg kg^{-1} i/v) then 0·25–0·75 mg kg^{-1} min^{-1})
Inhalation anaesthesia	Isoflurane using a face-mask	Isoflurane using a face-mask
Analgesia	Buprenorphine (0·01–0·05 mg kg^{-1} s/c or i/p, t.i.d.). Carprofen (5 mg kg^{-1} s/c, u.i.d.)	Buprenorphine (0·05–0·1 mg kg^{-1} s/c or i/p, t.i.d.)

	Gerbil	Hamster
Adult body weight	55–100 g	85–150 g
Body temperature	39°C	37·4°C
Respiration rate	90 min^{-1}	80 min^{-1}
Resting heart rate	260–300 min^{-1}	350 min^{-1}
Sedation/pre-medication	Fentanyl/fluanisone (Hypnorm) (0·5–1·0 ml kg^{-1} i/p)	Fentanyl/fluanisone (Hypnorm) (0·5 ml kg^{-1} i/p)
Injectable anaesthesia		
Short term (5–15 min)	–	–
Medium term (20–30 min)	Metomidate (50 mg kg^{-1}) plus fentanyl (0·05 mg kg^{-1}) s/c	Fentanyl/fluanisone and midazolam (4 ml kg^{-1} i/p of 2 parts water for injection, 1 part Hypnorm and 1 part midazolam)
Long term (1–12 h)	–	Use regimen for medium term, with additional fentanyl/fluanisone (0·5 ml kg^{-1} i/p) every 20–40 min and midazolam (5 mg kg^{-1} i/p) every 2–4 hours
Inhalational anaesthesia	Isoflurane using a face-mask	Isoflurane using a face-mask
Analgesia	Buprenorphine (0·01–0·05 mg kg^{-1} s/c or i/p, t.i.d.)	Buprenorphine (0·01–0·05 mg kg^{-1} s/c or i/p, t.i.d.)

	Guinea pig	Rabbit
Adult body weight	500–1000 g	3–6 kg
Body temperature	38°C	38°C
Respiration rate	120 min^{-1}	55 min^{-1}
Resting heart rate	155 min^{-1}	220 min^{-1}
Sedation/pre-medication	Fentanyl/fluanisone (Hypnorm) (1·0 ml kg^{-1} i/p)	Fentanyl/fluanisone (Hypnorm) (0·4 ml kg^{-1} i/p)
Injectable anaesthesia		
Short term (5–15 min)	Alphaxalone/alphadolone (10–12 mg kg^{-1} i/v)	Methohexitone (10–15 mg kg^{-1} i/v, 1% solution)

	Guinea pig	Rabbit
Medium term (20–30 min)	Fentanyl/fluanisone and midazolam (8 ml kg^{-1} i/p of 2 parts water for injection, 1 part Hypnorm and 1 part midazolam)	Fentanyl/fluanisone (0.3 ml kg^{-1} i/m) and midazolam (0·5–2 mg kg^{-1} i/v)
Long-term (1–12 h)	Continue medium-term regimen with fentanyl/ fluanisone (1·0 ml kg^{-1} h^{-1} i/p) and midazolam (2 mg kg^{-1} i/p, every 4 hours)	Continue medium-term regimen with fentanyl/ fluanisone (0·1–0·3 ml kg^{-1} h^{-1} i/v (preferably dilute 1: 10 and infuse 1–3 ml kg^{-1} h^{-1}, or use fentanyl alone, 0·5–2 ug kg^{-1} min^{-1}) and midazolam (1 mg kg^{-1} i/v, every 4 hours)
Inhalational anaesthesia	Isoflurane using a face-mask	Isoflurane using a Bain circuit or Ayre's T-piece
Analgesia	Buprenorphine (0·01–0·05 mg kg^{-1} s/c or i/p, t.i.d.).	Buprenorphine (0·05–0·1 mg kg^{-1} s/c or i/p, t.i.d.). Carprofen mg kg^{-1}. 1·5 mg kg^{-1}, per os, b.i.d.

	Dog	Cat
Adult body weight	15–20 kg	3–5 kg
Body temperature	38·3°C	38·6°C
Respiration rate	25 min^{-1}	26 min^{-1}
Resting heart rate	100 min^{-1}	150 min^{-1}
Sedation/pre-medication	Medetomidine (10–80 μg kg^{-1} i/m or s/c)—reverse with atipamezole (50–400 μg kg^{-1} i/v or i/m) if necessary)	Ketamine (5–20 mg kg^{-1} i/m). Medetomidine (50–150 μg kg^{-1} s/c or i/m)—reverse with atipamezole (0·2 μ mg kg^{-1} i/v, 0·6 mg kg^{-1} i/m) if necessary
Injectable anaesthesia		
Short term (5–15 min)	Propofol (5–7·5 mg kg^{-1} i/v)	Propofol (5–8 mg kg^{-1} i/v)
Medium term (20–30 min)	Continue propofol (0·2–0·4 mg kg^{-1} min^{-1} i/v)	Continue propofol (0·2–0·3 mg kg^{-1} min^{-1} i/v) or ketamine (7 mg kg^{-1} i/m) plus medetomidine (80 μg kg^{-1} i/m or s/c)

	Dog	Cat
Long term (1–12 h)	Propofol (0·2–0·4 mg kg^{-1} min^{-1}	Propofol (0·2–0·3 mg kg^{-1} min^{-1})
Inhalational anaesthesia	Isoflurane using a Bain or Magill circuit	Isoflurane using an Ayre's T-piece or Bain circuit
Analgesia	Buprenorphine (0.005–0.02 mg kg^{-1} s/c or i/p, t.i.d.) or carprofen (4 mg kg^{-1} s/c, u.i.d.)	Buprenorphine (0.005–0.01 mg kg^{-1} s/c or i/p, t.i.d.) or carprofen (4 mg kg^{-1} s/c, u.i.d.)

	Ferret	Pig
Adult body weight	500–1000 g	40–200 kg
Body temperature	39°C	39°C
Respiration rate	33–36 min^{-1}	18–12 min^{-1}
Resting heart rate	250 min^{-1}	220 min^{-1}
Sedation/pre-medication	Ketamine (20–30 mg kg^{-1} i/m)	Ketamine (10–15 mg kg^{-1} i/m) plus diazepam or midazolam (1–2 mg kg^{-1} i/m)
Injectable anaesthesia		
Short term (5–15 min)	Alphaxalone/alphadolone (8–12 mg kg^{-1} i/v)	Propofol (2·5–3·5 mg kg^{-1} i/v)
Medium term (20–30 min)	Ketamine (25 mg kg^{-1} i/m) and xylazine (12 mg kg^{-1} i/m)	Propofol, as above, then 0·1–0·2 mg kg^{-1} min^{-1} (preferable to add alfentanil 2–5 μg kg^{-1} min^{-1} with IPPV)
Long term (1–12 h)	Use inhalational agents	Continue medium-term regimen as above
Inhalational anaesthesia	Isoflurane using a face-mask	Isoflurane using a Bain or Magill circuit or circle system
Analgesia	Buprenorphine (0·01–0·05 mg kg^{-1} s/c or i/p, t.i.d.). Flunixin 0·5–2 mg kg s/c, u.i.d.	Buprenorphine (0·05–0·1 mg kg^{-1} s/c or i/p, t.i.d.). Carprofen 2–4 mg kg i/v or s/c, u.i.d.

	Sheep	Goat
Adult body weight	60–80 kg	40–100 kg
Body temperature	39·1°C	39·4°C
Respiration rate	20 min^{-1}	20 min^{-1}
Resting heart rate	75 min^{-1}	80 min^{-1}
Sedation/pre-medication	Xylazine (0·2 mg kg^{-1} i/m)	Xylazine (0·05 mg kg^{-1} i/m)
Injectable anaesthesia		
Short term (5–15 min)	Propofol (4–5 mg kg^{-1} i/v)	Propofol (4–5 mg kg^{-1} i/v)
Medium term (20–30 min)	Ketamine (10–15 mg kg^{-1} i/m or 4 mg kg^{-1} i/v) and xylazine (1 mg kg^{-1} i/v)	Ketamine (5–10 mg kg^{-1} i/v) plus diazepam (0·5–1 mg kg^{-1} i/v)
Long term (1–12 h)	Alphaxalone/alphadolone 2–3 mg kg^{-1} i/v then 0·1–0·2 mg kg^{-1} min^{-1}	Alphaxalone/alphadolone 2–3 mg kg^{-1} i/v then 0.1–0·2 mg kg^{-1} min^{-1}
Inhalational anaesthesia	Isoflurane using a Bain or Magill circuit or circle system	Isoflurane using a Bain or Magill circuit or circle system
Analgesia	Buprenorphine (0·005–0.01 mg kg^{-1} i/m or i/v, q.i.d.). Flunixin 2 mg kg s/c or i/v u.i.d.	Buprenorphine (0·005–0.01 mg kg^{-1} i/m or i/v, q.i.d.). Flunixin 2 mg kg^{-1} u.i.d.

	Primate (marmoset)	Primate (rhesus)
Adult body weight	500 g	8–12 kg
Body temperature	38·5–40°C	39°C
Respiration rate	–	35 min^{-1}
Resting heart rate	225 min^{-1}	150 min^{-1}
Sedation/pre-medication	Alphaxalone/alphadolone (12–18 mg kg^{-1} i/m)	Ketamine (5–25 mg kg^{-1} i/m)
Injectable anaesthesia		
Short term (5–15 min)	Alphaxalone/alphadolone (10–12 mg kg^{-1} i/v)	Propofol (7–8 mg kg^{-1} i/v) (NB, 2–3 mg kg^{-1} when ketamine used as pre-med)
Medium term (20–30 min)	Continue alphaxalone/ alphadolone at 0·3–0·6 mg kg^{-1} i/v	Ketamine (10 mg kg^{-1} i/m) plus xylazine (0·5 mg kg^{-1} i/m)
Long term (1–12 h)	Propofol, 7–8 mg kg i/v, then 0·2–0·5 mg kg^{-1} min i/v.	Continue Propofol, 0·2–0·5 mg kg^{-1} min^{-1} i/v

	Primate (marmoset)	Primate (rhesus)
Inhalational anaesthesia	Isoflurane using a Bain circuit or Ayre's T-piece	Isoflurane using a Bain circuit, Ayre's T-piece (< 10 kg) or Magill circuit (> 10 kg)
Analgesia	Buprenorphine (0·005–0·01 mg kg^{-1} i/m or i/v, q.i.d.). Flunixin 2 mg kg^{-1} s/c, u.i.d.	Buprenorphine (0·005–0·01 mg kg^{-1} i/m or i/v, t.i.d.). Flunixin 2 mg kg^{-1} u.i.d.

APPENDIX 2

Estimation of Required Quantities of Volatile Anaesthetic Agents and Anaesthetic Gases

Oxygen

Oxygen cylinders are coloured black with a white top segment in the UK and green in the USA. A size E cylinder when full contains approximately 680 litres of oxygen, sufficient for 340 hours use at 2 l min^{-1}. The quantity of gas remaining (in litres) in a size E cylinder can be estimated from the pressure, in psi, multiplied by 0.3.

Nitrous oxide

Nitrous oxide cylinders (coloured blue in the UK and in the USA) contain liquid N_2O, and the pressure reading on the pressure reducing valve does not indicate whether the cylinder is full or almost empty. When the pressure does fall, it will do so very rapidly as the cylinder empties. A full size E cylinder of nitrous oxide can deliver approximately 1800 litres of gas (at room temperature), in other words 900 hours at 2 l min^{-1}.

Volatile anaesthetic agents

The quantity of volatile anaesthetic required can be calculated from the molecular weight (1 gram mole of anaesthetic produces 22.4 litres of vapour at standard temperature and pressure) and the density of the liquid anaesthetic. For the most commonly used anaesthetics, at a temperature of 21°C, data are given in Table 1.

TABLE 1

Agent	Liquid density ($g\ ml^{-1}$)	Molecular weight	Volume from 1 ml (litres)	Concentration for maintenance	Quantity of agent ($ml\ min^{-1}$) for maintenance at $4\ l\ min^{-1}$ fresh gas flow
Enflurane	1·52	184·5	0·198	2%	0·4
Halothane	1·87	197	0·228	1·5%	0·26
Isoflurane	1·5	184·5	0·195	2%	0·41
Methoxyflurane	1·43	146	0·235	0·4%	0·07

From this, and a knowledge of the price per bottle, the costs of the anaesthetic agents can be compared. At the time of publication, typical costs of the agents were as shown in Table 2.

TABLE 2

Agent	Cost per unit	Cost per ml	Cost per minute with $4\ l\ min^{-1}$ flow (see Table 1)
Enflurane	£49.50 per 250 ml	£0.198	£0.079
Halothane	£10.50 per 250 ml	£0.042	£0.011
Isoflurane	£41 per 100 ml	£0.41	£0.17
Methoxyflurane	£73.3 per 100 ml	£0.73	£0.05

APPENDIX 3

Examples of Dilutions of Anaesthetic Mixtures for Small Rodents

- Look up dose of each drug in mg kg^{-1}.
- Convert to ml kg^{-1} according to concentration of stock solution.
- Convert to ml per 100 g (rats), ml per 10 g (mice).
- Add diluent (water for injection, WFI) to make a sensible volume per animal (e.g. 0·2 ml per 100 g for rats, 0·1 ml per 10 g for mice).
- **Before making up these cocktails confirm that the strengths of the stock solutions you are using are the same as those used here.**

RAT

Except for fentanyl/medetomidine and fentanyl/fluanisone/midazolam and the reversal agents, the volumes listed below make up sufficient material for animals with a total body weight of 1 kg (i.e. four or five young adults). The dilutions are adjusted to provide a volume of injectate of 0·2 ml per 100 g.

Ketamine/xylazine

75 mg kg^{-1} ketamine + 10 mg kg^{-1} xylazine i/p
0·75 ml (75 mg) ketamine + 0·5 ml (10 mg) xylazine + 0·75 ml WFI gives 4–5 doses of 0·2 ml per 100 g

Ketamine/medetomidine

75 mg kg^{-1} ketamine + 0·5 mg kg^{-1} medetomidine i/p
0·75 ml (75 mg) ketamine + 0·5 ml (0·5 mg) medetomidine + 0·75 ml WFI gives 4–5 doses of 0·2 ml per 100 g

Ketamine/midazolam

75 mg kg^{-1} ketamine + 5 mg kg^{-1} midazolam i/p
0·75 ml (75 mg) ketamine + 1 ml (5 mg) midazolam + 0·25 ml WFI gives 4–5 doses of 0·2 ml per 100 g

Ketamine/acepromazine

75 mg kg^{-1} ketamine + 2·5 mg kg^{-1} acepromazine i/p
0·75 ml (75 mg) ketamine + 0·25 ml (2·5 mg) acepromazine + 1 ml WFI gives 4–5 doses of 0·2 ml per 100 g

Fentanyl/fluanisone (Hypnorm)/midazolam

1 ml Hypnorm (0·315 mg fentanyl per ml; 10 mg fluanisone per ml) + 1 ml midazolam (5 mg) + 2 ml WFI gives 4–5 doses of 0·27 ml per 100 g i/p (add water for injection to Hypnorm *before* adding midazolam)

Fentanyl/medetomidine

300 μg kg^{-1} fentanyl + 300 μg kg^{-1} medetomidine i/p
2 ml (100 μg) fentanyl + 0·1 ml (100 μg) medetomidine gives 1 dose of 0·63 ml per 100 g

Reversal for fentanyl/medetomidine

Nalbuphine 2 mg kg^{-1} s/c
0·2 ml (2 mg) + 0·8 ml WFI gives 4–5 doses of 0·1 ml per 100 g

Atipamezole 1 mg kg^{-1} s/c
0·2 ml (1 mg) + 0·8 ml WFI gives 4–5 doses of 0·1 ml per 100 g

MOUSE

The quantities listed below make up sufficient material for animals with a total body weight of 500 g (i.e. 15–20 young adults). Except for Hypnorm/midazolam, the dilution is adjusted to provide a volume of injectate of 0·1 ml per 10 g.

Ketamine/xylazine

100 mg kg^{-1} ketamine + 10 mg kg^{-1} xylazine i/p
0·5 ml (50 mg) ketamine + 0·25 ml (5 mg) xylazine + 4·25 ml WFI gives about 17 doses of 0·1 ml per 10 g

Ketamine/midazolam

100 mg kg^{-1} ketamine + 5 mg kg^{-1} midazolam i/p
0·5 ml (50 mg) ketamine + 0·5 ml (2.5 mg) midazolam + 4·0 ml WFI gives about 17 doses of 0·1 ml per 10 g

Ketamine/acepromazine

100 mg kg^{-1} ketamine + 5 mg kg^{-1} acepromazine i/p
0·5 ml (50 mg) ketamine + 1·25 ml (2·5 mg) acepromazine + 3·25 ml WFI gives about 17 doses of 0·1 ml per 10 g

Fentanyl/fluanisone (Hypnorm)/midazolam

1 ml Hypnorm (0·315 mg fentanyl per ml; 10 mg fluanisone per ml) + 1 ml midazolam (5 mg) + 2 ml WFI gives 4–5 doses of 0·1 ml per 10 g i/p

Ketamine/medetomidine

75 mg kg^{-1} ketamine + 1 mg kg^{-1} medetomidine
0·38 ml (38 mg) ketamine + 0·5 ml (0·5 mg) medetomidine + 4·12 ml WFI gives about 17 doses of 0·1 ml per 10 g

GUINEA PIG

The quantities listed below make up sufficient material for animals with a total body weight of 1 kg (i.e. two young adults). Except for ketamine/acepromazine and Hypnorm/midazolam, the dilution is adjusted to provide a volume of injectate of 2 ml kg^{-1}

Ketamine/xylazine

40 mg kg^{-1} ketamine + 5 mg kg^{-1} xylazine i/p
0·4 ml (40 mg) ketamine + 0·25 ml (5 mg) xylazine + 1·35 ml WFI gives enough for 1 kg at 2·0 ml kg^{-1}

Ketamine/acepromazine

125 mg kg^{-1} ketamine + 5 mg kg^{-1} acepromazine i/p
1·25 ml (125 mg) ketamine + 2·5 ml (5 mg) acepromazine + 0·25 ml WFI gives enough for 1 kg at 4·0 ml kg^{-1}

Fentanyl/fluanisone (Hypnorm)/midazolam

2 ml Hypnorm (0·315 mg fentanyl per ml; 10 mg fluanisone per ml + 2 ml midazolam (5 mg) + 4 ml WFI gives enough for 1 kg at 8 ml kg^{-1} i/p

Ketamine/medetomidine

40 mg kg^{-1} ketamine + 0·5 mg kg^{-1} medetomidine
0·4 ml (40 mg) ketamine + 0·5 ml medetomidine + 1·1 ml WFI gives enough for 1 kg at 2·0 ml kg^{-1}

BIRDS

Equithesin

5·25 g Chloral hydrate + 12·5 ml absolute alcohol
20·25 ml Sagatal (pentobarbitone, 60 mg ml)
49·5 ml propylene glycol
2·65 g magnesium sulphate + 25 ml sterile water for injection

Mix all of these, and make up total volume to 125 ml with water for injection. Dose rate 2·5 ml/kg i/m.

APPENDIX 4

Basic Equipment for Anaesthesia of Laboratory Animals

Anaesthetic trolley with nitrous oxide and oxygen supply
or
Gas regulator and flowmeter, oxygen cylinder.

Calibrated vaporizer (for isoflurane, halothane or methoxyflurane).

Antistatic tubing (6 mm) for connection of small face-masks, circuits, etc. to fresh gas output.

Gas-scavenging unit, if volatile anaesthetics are to be used.

Bain coaxial circuit.

Paediatric T-piece and Oxford endotracheal tube connectors.

Electric clippers, curved scissors (15 cm Mayo type).

Linen tape (1·25 cm) for anchoring tubes and other equipment to animal and to operating table.

Adhesive bandage (2·5 cm) for anchoring catheters/needles in vein.

Disposable needles, 26 G to 19 G.

Over-the-needle catheters, 23 G to 19 G.

Disposable syringes, 0·5 ml to 50 ml.

Intravenous infusion sets, preferably paediatric size burette.

Endotracheal tubes, 2·5 mm to 4·5 mm (uncuffed); 5 mm to 12 mm (cuffed).

Introducer for endotracheal tube.

Standard-sized endotracheal tube connectors.

Laryngoscope handle and blades (see Table 3.1 for style and sizes).

APPENDIX 5

Anaesthetic Drugs — UK and USA generic names, trade names and manufacturers

UK approved name	US approved name	UK trade name	Manufacturer	US trade name	Manufacturer
Acepromazine maleate	Acepromazine maleate	ACP	C-Vet VP	Acepromazine	Fermenta Animal Health
Adrenaline	Epinephrine	–	Mallinckrodt	–	Mallinckrodt Elkins-Sinn
Alpha-chloralose	Alpha-chloralose	–	BDH	–	Fisher Scientific
Alphaxalone/ alphadolone	Alphaxalone/ alphadolone	Saffan	Mallinckrodt	–	–
Alcuronium	–	Alloferin	Roche	–	–
Alfentanil	–	Rapifen	Janssen	–	–
Altracurium	–	Tracrium	Wellcome	–	–
Atropine sulphate	Atropine sulphate	Atrocare	Animalcare C-Vet VP	–	Fort Dodge
Azaperone	Azaperone	Stresnil	Janssen	Stresnil	Mallinckrodt
Bupivacaine hydrochloride	Bupivacaine hydrochloride	Marcain	Astra	Marcaine Sensorcaine	Winthrop
Buprenorphine	Buprenorphine	Temgesic Vetergesic	Reckitt & Colman Animalcare	–	–
Butorphanol	Butorphanol Torbutrol	Torbugesic Torbutol	Willows	Torbugesic	Bristol-Myers Squibb

Anaesthetic Drugs — *continued*

UK approved name	US approved name	UK trade name	Manufacturer	US trade name	Manufacturer
Carprofen	–	Zenecarp	C-Vet VP	–	–
Dextro-propoxyphene	Propoxyphene	Dextro-propoxyphene Doloxene	– Lilly	–	–
Diazepam	Diazepam	Valium Diazemuls	Roche Dumex	Valium	Roche
Dihydrocodeine	–	DF-118	Glaxo	–	–
Diprenorphine	Diprenorphine	Revivon	C-Vet VP	M50–50	Lemmon
Doxapram-hydrochloride	Doxapram-hydrochloride	Dopram-V	Willows	Dopram-V	Whitehall Robbins
Droperidol	Droperidol	Droleptan	Janssen	Inapsine	–
Enflurane	Enflurane	Enflurane	Abbott	Ethrane	Abbott
Etomidate	Etomidate	Hypnomidate	Janssen	Amidate	Abbott
Etorphine	Etorphine	–	–	M-99	American Cyanamid
Etorphine–acepromazine	–	Immobilon LA	C-Vet VP	–	–
Etorphine–metho-trimeprazine	–	Immobilon SA	C-Vet VP	–	–
Fentanyl	Fentanyl	Sublimaze	Janssen	Sublimaze	–
Fentanyl–droperidol	Fentanyl–droperidol	Thalamonal	Janssen	Innovar Vet	Mallinckrodt

Anaesthetic Drugs — *continued*

UK approved name	US approved name	UK trade name	Manufacturer	US trade name	Manufacturer
Fentanyl–fluanisone	–	Hypnorm	Janssen	–	–
Flumazenil		Anexate	Roche	–	–
Flunixin	Flunixin	Friadyne	Schering-Plough Animal Health	–	–
Gallamine	Gallamine	Flaxedil	Rhône-Poulenc	Flaxedil	American Cyanamid
Halothane	Halothane	Fluothane Halothane	Mallinckrodt Rhône Mérieux	Fluothane	Fort Dodge
Inactin	Inactin	–	Semat	–	RBI
Isoflurane	Isoflurane	Isoflurane	Abbott	Forane	Ohmeda
Ketamine	Ketamine	Vetalar Ketaset	Parke-Davis Willows	Vetalar Ketaset	Parke-Davis Bristol-Myers Squibb
Lignocaine	Lidocaine	Xylocaine	Astra (UK)	Xylocaine	Astra USA
Medetomidine	–	Domitor	SmithKline Beecham	–	–
Methadone	Methadone	Physeptone	Wellcome	Dolophine	Eli Lilly
Methohexitone	Methohexital	Brietal	Elanco	Brevital	Eli Lilly
Methoxyflurane	Methy-oxyflurane	Metofane	C-Vet VP	Penthrane	Abbott
Metomidate	Metomidate	Hypnodil	Janssen	–	–

Anaesthetic Drugs — *continued*

UK approved name	US approved name	UK trade name	Manufacturer	US trade name	Manufacturer
Midazolam	Midazolam	Hypnovel	Roche	Versed	Roche
Morphine	Morphine	Duromorph Oramorph SR MST Continus	LAB Boehringer Ingelheim Napp	–	–
Naloxone	Naloxone	Narcan	Du Pont	Narcan	Endo
Oxymorphone	Oxymorphone	–	–	Numorphan	Endo
Pancuronium	Pancuronium	Pavulon	Organon Teknika	Pavulon	Organon
Paracetamol	Acetominophen	–	–	–	–
Paracetamol/ codeine	–	Pardale	Arnolds	–	–
Pentazocine	Pentazocine	Fortral	Sanofi Winthrop	Talwin-V	Winthrop
Pentobarbitone	Pentobarbital	Sagatal	Rhône Mérieux	Nembutal	Abbott
Pethidine	Meperidine	Pethidine	Arnolds	Demerol	Elkins-Sinn
Propofol	–	Diprivan Rapinovet	Zeneca Mallinckrodt	–	–
Suxamethonium	Succinylcholine	Anectine Scoline	Wellcome Evans	Sucostrin	Bristol-Myers Squibb

Anaesthetic Drugs — *continued*

UK approved name	US approved name	UK trade name	Manufacturer	US trade name	Manufacturer
Thiopentone	Thiopental Thiovet	Intraval C-Vet VP	Rhône Mérieux	Pentothal	Abbott
Tiletamine/ Zolezepam	Tiletamine/ Zolezepam	Zoletil	Virbac	Telazol	Parke Davis
Trichloroethylene	Trichloroethylene	Trilene	Zeneca	Trilene	Ohmeda
Urethane	Urethane	–	–	Urethane	Merck
Vecuronium	–	Norcuron	Organon-Teknika	–	–
Xylazine	–	Rompun	Bayer	Rompun	Bayer Animal Health

APPENDIX 6

Addresses of Drug Manufacturers and Suppliers

Abbott Laboratories
100 Abbott Park Road
Abbott Park, IL 60064
USA
(Tel: 708 937 6100; fax: 708 937 1511)

Abbott Laboratories Ltd
Abbott House
Norden Road
Maidenhead
Berks SL6 4BW
UK
(Tel: 01628 773355; fax: 01628 773803)

American Cyanamid Co.
PO Box 400
Princeton, NJ 08543–0400
USA
(Tel: 609 799 0400; fax: 609 275 5235)

Animalcare Ltd
Common Road
Dunnington
York YO1 5RU
UK
(Tel: 01904 488661; fax: 01904 488534)

Arnolds Veterinary Products Ltd
Cartmel Drive
Harlescott
Shrewsbury
Shropshire SY1 3TB
UK
(Tel: 01743 231632; fax: 01743 352111)

Astra Pharmaceuticals Ltd
Home Park
King's Langley
Herts WD4 8DH
UK
(Tel: 01923 266191; fax: 01923 270529)

Astra USA, Inc.
50 Otis Street
Westborough, MA 01581
USA
(Tel: 508 366 1100; fax: 508 366 7406)

Bayer Animal Health
12707 Shawnee Mission Parkway
PO Box 390
Shawnee, KS 66216
USA
(Tel: 800 255 6517 or 800 633 0805)

Bayer (UK) Ltd
Veterinary Business Group
Eastern Way
Bury St Edmunds
Suffolk IP32 7AH
UK
(Tel: 01284 763299; fax: 01284 702810)

BDH Ltd
Broom Road
Poole
Dorset BH12 4NN
UK
(Tel: 01202 745520)

Boehringer Ingelheim Ltd
Ellesfield Avenue
Bracknell
Berkshire, RG12 4YS
UK
(Tel: 01344 424600)

Bristol-Myers Squibb
Division of Solvay Animal Health Inc.
PO Box 4755
Syracuse, NY 13221–4755
USA
(Tel: 315 432 2000)

C-Vet Veterinary Products
Grampian Pharmaceuticals Ltd
Marathon Place
Moss Side Industrial Estate
Leyland
Lancs PR5 3QN
UK
(Tel: 01772 452421)

Dumex Ltd
Longwick Road
Princes Risborough
Aylesbury
Bucks HP17 9UZ
UK
(Tel: 01844 274414; fax: 01844 274420)

Duncan Flockhart & Co. Ltd
700 Oldfield Lane North
Greenford
Middlesex UB60 0HD
UK

Du Pont Pharmaceuticals Ltd
Avenue One
Letchworth Garden City
Stevenage
Herts SG6 2HU
UK
(Tel: 01462 488200; fax: 01462 488245)

Elanco Products Ltd
Dextra Court
Chapel Hill
Basingstoke
Hants RG21 2SY
UK
(Tel: 01256 53131; fax: 01256 485081)

Eli Lilly & Co.
Dextra Court
Chapel Hill Basingstoke
Hants RG21 2SY
UK
(Tel: 01256 473241; fax: 01256 485246)

Elkins-Sinn, Inc.
Subsidiary of Wyeth-Ayerst
2 Esterbrook Lane
Cherry Hill, NJ 08003–4099
USA
(Tel: 609 424 3700)

Endo Laboratories Inc.
4301 Lancaster Pike
Barley Mill Plaza
Building 23
Wilmington, DE 19880
USA
(Tel: 302 892 1841 or 302 992 3006)

Evans Medical Ltd
Langhurst
Horsham
Sussex RH12 4QD
UK
(Tel: 01403 41400; fax: 01403 61101)

Fermenta Animal Health
TechAmerica Veterinary Products
15th & Oak
PO Box 338
Elwood, KS 66024
USA
(Tel: 913 365 5288)

Fisher Scientific Co.
711 Forbes Avenue
Pittsburgh, PA 15219
USA
(Tel: 412 562 8300; fax: 800 926 1166)

Fort Dodge Laboratories
800 Fifth Street NW
Fort Dodge IA 50501
USA
(Tel: 515 955 4600; fax: 515 955 9191)

Glaxo Laboratories Ltd
Stockley Park West
Uxbridge
Middlesex UB11 1BT
UK
(Tel: 0181 990 9444)

Janssen Animal Health
Grove
Wantage
Oxon OX12 0DQ
UK
(Tel: 01235 777555; fax: 01235 777444)

Janssen Pharmaceutica
Turnhoutsebaan 30
2340 Beerse
Belgium

LAB (Laboratories for Applied Biology Ltd)
91 Amherst Park
London N16 5DR
UK
(Tel: 0181 800 2252; fax: 0181 809 6884)

Laboratorios Virbac SA
Angel Guimera 179–181
Esplugues de Llabregat
Barcelona
Spain

Lemmon Co.
650 Cathill Road
Sellersville, PA 18960
USA
(Tel: 800 545 8800 or 215 256 7855)

Mallinckrodt Veterinary Ltd
Breakspear Road South
Harefield
Uxbridge
Middlesex UB9 6LS
UK
(Tel: 01895 626355; fax: 01895 626444)

Mallinckrodt Veterinary Inc.
421 East Hawley Street
Mundelain, IL 60060
USA
(Tel: 800 525 9480; fax: 800 462 3720)

Merck & Co.
126 East Lincoln Avenue
PO Box 2000
Rahway, NJ 07065
USA
(Tel: 908 594 4000)

Merck Sharp & Dohme Ltd
Hertford Road
Hoddesdon
Herts EN11 9BU
UK
(Tel: 01992 467272; fax: 01992 451059)

Napp Laboratories Ltd
Cambridge Science Park
Milton Road
Cambridge CB4 4BH
UK
(Tel: 01223 424444)

Ohmeda
Ohmeda Drive
PO Box 7550
Madison,
WI 53707–7550
USA
(Tel: 608 221 1551; fax: 608 221 4384)

Organon Teknika Ltd
Cambridge Science Park
Milton Road
Cambridge CB4 4FL
UK
(Tel: 01223 423650; fax: 01223 420264)

Parke-Davis Veterinary
Usk Road
Pontypool
Gwent NR4 0YH
UK
(Tel: 01495 762468; fax: 01495 762628)

Parke-Davis
Division of Warner Lambert Co.
201 Tabor Road
Morris Plains, NJ 07950
USA
(Tel: 201 540 2000 or 800 223 0432)

RBI (Research Biomedicals International)
One Strathmore Road
Natick, MA 01760–2447
USA
(Tel: 800 736 3690)

Reckitt & Colman
Dansom Lane
Hull
N. Humberside HU8 7DS
UK
(Tel: 01482 26151; fax: 01482 25322)

Rhône Mérieux Ltd
Spire Green Centre
Harlow
Essex CM19 5TS
UK
(Tel: 01279 439444; fax: 01279 426556)

Rhône Poulenc Rorer Ltd
RPR House
St Leonards Road
Eastbourne
East Sussex BN21 3YG
UK
(Tel: 01323 21422; fax: 01323 20666)

Roche Laboratories
Division of Hoffman-La Roche Inc.
340 Kingsland Road
Nutley, NJ 07110
USA
(Tel: 201 235 5000)

Roche Products Ltd
Vitamin & Chemical Division
Broadwater Road
Welwyn Garden City
Herts AL7 3AY
UK
(Tel: 01707 366000; fax: 01707 329587)

Sanofi Winthrop Ltd
1 Onslow Street
Guildford
Surrey GU1 4YS
UK
(Tel: 01483 505515; fax: 01483 354332)

Schering-Plough Animal Health
Schering-Plough House
Shire Park
Welwyn Garden City
Herts AL7 1TW
UK
(Tel: 01707 363636; fax: 01707 363690)

Semat Technical (UK) Ltd
One Executive Park
Hatfield Road
St Albans
Herts AL1 4TA
UK
(Tel: 01727 841414; fax: 01727 843965)

SmithKline Beecham
Hunters Chase
Walton Oaks
Dorking Road
Tadworth
Surrey KT20 7NT
UK
(Tel: 01737 364700; fax: 01737 325600)

Squibb Pharmaceutical Co.
ER Squibb & Sons Inc.
PO Box 4000
Princeton, NJ 08540–4000
USA
(Tel: 609 252 4000)

Wellcome Medical Division
The Wellcome Foundation Ltd
Crewe Hall
Crewe
Cheshire CW1 1UB
UK
(Tel: 01270 583151; fax: 01270 589305)

Whitehall Robbins
1407 Cummings Drive
Richmond, VA 23220
USA
(Tel: 804 257 2000)

Willows Francis Ltd
3 Charlwood Court
County Oak Way
Crawley
West Sussex RH11 7XA
UK
(Tel: 01293 614141; fax: 01293 614324)

Winthrop Pharmaceuticals
90 Park Avenue
New York, NY 10016
USA
(Tel: 212 907 2000)

Zeneca Pharmaceuticals
Alderley House
Alderley Park
Macclesfield
Cheshire SK10 4TF
UK
(Tel: 01625 582828)

APPENDIX 7

Sources of Anaesthetic Apparatus and Related Equipment

Equipment	Manufacturers or Distributors (numbers refer to addresses on next page)
Anaesthetic circuits and anaesthetic machines	3, 8, 11, 20, 21, 22, 24, 33, 34
Endotracheal tubes, introducers	8, 18, 25, 33, 35, 36
Laryngoscopes, mouth gags	3, 5, 18, 21, 22, 24, 27, 32, 33
Catheters	1, 8, 15, 25, 33, 36
Infusion pumps	10, 12, 13, 16
Incubators	4, 12, 13, 31, 35
Homeothermic blankets	12, 13, 23
Gas-scavenging apparatus	15, 18, 21, 22, 24, 35
Monitoring equipment	4, 6, 7, 12, 13, 17, 18, 21, 22, 23, 24, 26, 27, 28, 29, 30
Respiratory monitors	10, 15, 17, 29, 30, 33, 34
Thermometers	2, 15, 17, 19, 33
Ventilators	3, 11, 12, 13, 14, 18, 20, 21, 22, 23, 24
Wright's respirometer	9, 18

Addresses of manufacturers and distributors of anaesthetic apparatus and related equipment

1. Abbott Laboratories
   Queenborough
   Kent ME11 5EL
   UK
   (Tel: 10795 580099 or 580303)

2. Analog & Numeric Devices Ltd
   Swannington Road
   Broughton Astley
   Leicestershire LE9 6RD
   UK
   (Tel: 01455 285647;
   fax: 01455 285409)

3. Blease Medical Equipment Ltd
   Deansway
   Chesham
   Bucks HP5 2NX
   UK
   (Tel: 10494 784422)

4. Braintree Scientific, Inc.
   PO Box 361
   Braintree, MA 02184
   USA
   (Tel: 617 843 2202;
   fax 617 843 7932)

5. Brookwick, Ward & Co. Ltd
   88 Westlaw Place
   Whitehill Estate
   Glenrothes
   Fife KY6 2RZ
   UK
   (Tel: 01592 630052;
   fax: 01592 630109)

6. Burtons of Maidstone Ltd
   Unit 2 Guardian Industrial Estate
   Pattenden Lane
   Marden
   Kent TN12 9QJ
   UK
   (Tel: 01622 832929;
   fax 01622 832949)

7. Columbus Instruments
   950 North Hague
   PO Box 44049
   Columbus, OH 43204
   USA
   (Tel: 614 276 0861
   or 800 669 5011;
   fax: 614 276 0529)

8. Cox Surgical
   Edward Road
   Coulsdon
   Surrey CR5 2XA
   UK
   (Tel: 0181 668 2131;
   fax: 0181 668 4196)

9. Ferraris Medical Ltd
   26 Lea Valley Trading Estate
   Angel Road
   Edmonton
   London N18 3JD
   UK
   (Tel: 0181 807 3636)

10. Graseby Medical
    Colonial Way
    Watford
    Herts WD2 4LG
    UK
    (Tel: 01923 246434;
    fax: 01923 240273)

11. Hallowell Engineering & Manufacturing Corporation
    74 North Street
    Pittsfield, MA 01201
    USA
    (Tel: 413 445 4263)

12. Harvard Apparatus Ltd
    Fircroft Way
    Edenbridge
    Kent TN8 6HE
    UK
    (Tel: 0172 864001;
    fax: 01732 863356)
13. Harvard Apparatus Inc.
    22 Pleasant Street
    South Natick, MA 01760
    USA
    (Tel: 508 655 7000 or 800 272 2775;
    fax: 508 655 6029)
14. Hillmoore Electronic Consultants,
    22 High Street
    Hanslope
    Milton Keynes MK19 7LO
    UK
    (Tel/fax: 01908 511482)
15. International Market Supply
    Dane Mill
    Broadhurst Lane
    Congleton
    Cheshire CW12 1LA
    UK
    (Tel: 01260 275469;
    fax: 01260 276007)
16. IVAC Ltd
    IVAC House
    98–100 Bessborough Road
    Harrow
    Middlesex HA1 3DT
    UK
    (Tel: 0181 864 9624)
17. Kontron Instruments Ltd
    Blackmoor Lane
    Croxley Business Park
    Watford
    Herts WD1 8XQ
    UK
    (Tel: 01923 245991)
18. M&IE Dentsply
    Falcon Road
    Sowton Industrial Estate
    Exeter EX2 7NA
    UK
    (Tel: 01392 31331;
    fax: 01392 39927)
19. Miller and Weber, Inc
    1637 George Street
    Ridgewood
    Queens, NY 11385–5342
    USA
    (Tel: 718 821 7110;
    fax: 718 821 1673)
20. North American Drager
    148 B Quarry Road
    Telford, PA 18969
    USA
    (Tel: 215 723 9824;
    fax: 215 721 9561)
21. Ohmeda
    PO Box 75
    Hatfield
    Herts AL9 5JL
    UK
    (Tel: 01707 263570;
    fax: 01707 260065)
22. Ohmeda
    Ohmeda Drive
    PO Box 7550
    Madison, WI 53707–7550
    USA
    (Tel: 608 221 1551)
23. Palmer Bioscience
    Harbour Estate
    Sheerness
    Kent ME12 1RT
    UK
    (Tel: 01795 667551)

24. Penlon Ltd
Radley Road
Abingdon
Oxford OX14 3PH
UK
(Tel: 01235 554222;
fax: 01235 555252)

25. Portex Ltd
Hythe
Kent CT21 6JL
UK
(Tel: 01303 66863/60551;
fax: 01303 66761)

26. Sensor Devices Inc.
407 Pilot Court, #400 A
Waukesha, WI 53188
USA
(Tel: 414 524 1000;
fax: 414 524 1009)

27. Seward Medical
131 Great Suffolk Street
London SE1 1PP
UK
(Tel: 0171 928 9431)

28. Sharn Inc.
4801 George Road
Tampa, FL 33634–6200
USA
(Tel: 800 325 3671;
fax: 813 886 2701)

29. Silogic International Ltd
Unil L, Mill Green Business Park
Mill Green
Mitcham
Surrey CR4 4HT
UK
(Tel: 0181 646 2525;
fax: 0181 646 6622)

30. Silogic International Ltd
RR1 1544 Orchard Road
Stewartstown, PA 17363
USA
(Tel: 717 993 5093;
fax: 717 993 5289

31. Thermocare
PO Box 6069
Incline Village, NV 89450
USA
(Tel: 702 831 1201;
fax: 702 831 1230)

32. Timesco of London
Timesco House
1 Knights Road
London E16 2AT
UK
(Tel: 0171 511 1234;
fax: 0171 511 7888)

33. Veterinary Drug Co. plc
Derwent Valley Industrial Estate
Dunnington
York YO1 5RS
UK
(Tel: 01904 489958;
fax: 01904 488538)

34. Veterinary Instrumentation
62 Cemetery Road
Sheffield S11 8FP
UK
(Tel: 01742 70078;
fax: 01742 759471)

35. Viking Medical
PO Box 2142
Medford Lakes, NJ 08055
USA
(Tel: 800 920 1033 or 609 953 0138;
fax: 609 654 2417)

36. Vygon (UK) Ltd
Bridge Road
Cirencester
Glos. GL7 1PT
UK
(Tel: 01285 657051;
fax: 01285 650293)

INDEX